Lecture Notes in Computer Science   **12281**

Luís S. Barbosa · Mohammad Ali Abam (Eds.)

Topics in Theoretical Computer Science

Third IFIP WG 1.8 International Conference, TTCS 2020
Tehran, Iran, July 1–2, 2020
Proceedings

Editors
Luís S. Barbosa ⓘ
Universidade do Minho
Braga, Portugal

Mohammad Ali Abam ⓘ
Sharif University of Technology
Tehran, Iran

ISSN 0302-9743   ISSN 1611-3349   (electronic)
Lecture Notes in Computer Science
ISBN 978-3-030-57851-0   ISBN 978-3-030-57852-7   (eBook)
https://doi.org/10.1007/978-3-030-57852-7

LNCS Sublibrary: SL1 – Theoretical Computer Science and General Issues

This Springer imprint is published by the registered company Springer Nature Switzerland AG
The registered company address is: Gewerbestrasse 11, 6330 Cham, Switzerland

Preface

We are honored to bring you this collection of the revised selected papers from the third IFIP WG 1.8 International Conference on Topics in Theoretical Computer Science (TTCS 2020), which was held at the School of Computer Science, Institute for Research in Fundamental Sciences (IPM), Tehran, Iran, July 1–2, 2020.

TTCS is a bi-annual conference series, intending to serve as a forum for novel and high-quality research in all areas of Theoretical Computer Science. Two previous editions were held in 2015 and 2017, with proceedings also published by Springer in the *Lecture Notes in Computer Science* series, respectively as volumes 9541 and 10608. As before, the 2020 conference was held in cooperation with IFIP WG 1.8, on Foundations of Concurrency, and the European Association for Theoretical Computer Science.

TTCS 2020 received 24 submissions, which were carefully reviewed by the members of an International Program Committee comprising 45 leading scientists. At least three review reports were provided for each paper. After a participated final discussion inside the Program Committee eight papers were accepted, fixing the acceptance rate of the conference at approximately 34%.

Besides paper presentations, the program included two invited talks by the following world-renowned computer scientists:

- Filippo Bonchi, University of Pisa, Italy, on "Interacting Hopf Algebras: the theory of linear systems"
- Mohammad Taghi Hajiaghayi, University of Maryland, College Park, USA, on "Massively Parallel Algorithms and Applications for Maximal Matching and Edit Distance"

We would like to thank authors and invited speakers who contributed to the success of TTCS 2020. We are also grateful to all Program Committee members for their professional and hard work in providing expert review reports and thorough discussions leading to a very interesting and strong program. Last but not least, our gratitude extends to IPM for generously supporting the conference.

Due to the COVID-19 pandemic, TTCS 2020 was forced to run virtually, with live streaming of all talks and both synchronous and asynchronous interaction with participants. We acknowledge the excellent facilities provided by Sharif University of Technology, Iran, to hold the conference online.

July 2020

Luís S. Barbosa
Mohammad Ali Abam

Organization

Program Committee Chairs

Mohammad Ali Abam	Sharif University of Technology, Iran
Luís Soares Barbosa	Universidade do Minho, Portugal

Steering Committee

Farhad Arbab (Chair)	CWI Amsterdam and Leiden University, The Netherlands
Anuj Dawar	University of Cambridge, UK
Abbas Edalat	Imperial College, UK
Michael Fellows	University of Bergen, Norway
Wan Fokkink	Vrije Universiteit Amsterdam, The Netherlands
Hamid Sarbazi-azad	IPM and Sharif University, Iran

Program Committee

Track A: Algorithms and Complexity

Mohammad Ali Abam	Sharif University of Technology, Iran
Sepehr Assadi	Rutgers University, USA
Mohammad Hossein Bateni	Google Research, USA
Salman Abolfath Beigy	IPM, Iran
Hossein Esfandiari	Harvard University, USA
Omid Etesami	IPM, Iran
Marc van Kreveld	Utrecht University, The Netherlands
Mohammad Mahdian	Google Research, USA
Mohammad Mahmoody	University of Virginia, USA
Vahab Mirrokni	Google Research, USA
Gunter Rote	FU Berlin, Germany
Mohammadreza Salavatipour	University of Alberta, Canada
Masoud Seddighin	IPM, Iran
Saeed Seddighin	Harvard University, USA
Michiel Smid	Carleton University, Canada
Hamid Zarrabi-Zadeh	Sharif University of Technology, Iran

Track B: Logic, Semantics, and Programming Theory

Farhad Arbab	CWI, The Netherlands
Kyungmin Bae	Pohang University Science and Technology, South Korea

Christel Baier	Technische Universität Dresden, Germany
Luis Soares Barbosa	University of Minho, Portugal
Mário Benevides	Universidade Federal do Rio de Janeiro, Brazil
Simon Bliudze	Inria Lille, France
Filippo Bonchi	University of Pisa, Italy
Marcello Bonsangue	Leiden University, The Netherlands
Flavio Corradini	University of Camerino, Italy
Fredrik Dahlqvist	University College London, UK
Sergey Goncharov	FAU Erlangen-Nürnberg, Germany
Hossein Hojjat	Rochester Institute of Technology, USA
Mohammad Izadi	Sharif University of Technology, Iran
Sung-Shik Jongmans	Open University, The Netherlands
Alexander Knapp	University of Augsburg, Germany
Jan Kretinsky	Munich University of Technology, Germany
Alexandre Madeira	University of Aveiro, Portugal
Stefan Mitsch	Carnegie Mellon University, USA
Mohammad Reza Mousavi	University of Leicester, UK
Renato Neves	INESC TEC, Portugal
Peter Ölveczky	University of Oslo, Norway
Prakash Panangaden	McGill University, Canada
Elaine Pimentel	Universidade Federal do Rio Grande do Norte, Brazil
Subodh Sharma	IIT Delhi, India
Pawel Sobocinski	Taltech, Estonia
Ana Sokolova	University of Salzburg, Austria
Carolyn Talcott	Stanford University, USA
Benoit Valiron	LRI, France
Naijun Zhan	Chinese Academy of Science, China

Additional Reviewers

Mohammad Reza Bahrami	Shahin Kamali
Mohammad Sadegh Borouny	Eduard Kamburjan
Guillermina Cledou	Mehran Khodabandeh
Daniel Figueiredo	Mohamad Latifian
Ehsan Emamjomeh-Zadeh	Dave Parker
Claudio Gallicchio	Alireza Rezaei
Arnd Hartmanns	Mingji Xia

Contents

Contents

Dirac-Based Reduction Techniques for Quantitative Analysis of Discrete-Time Markov Models

Mohammadsadegh Mohagheghi$^{(\boxtimes)}$ and Behrang Chaboki

Vali-e-Asr University of Rafsanjan, Rafsanjan, Iran
{mohagheghi,b.chaboki}@vru.ac.ir
https://math.vru.ac.ir/computer-science

Abstract. Iterative methods are widely used in probabilistic model checking to compute quantitative reachability values. Redundant computation is a significant drawback of iterative methods that affects the performance of probabilistic model checking. In this paper we propose a new approach to avoid redundant computations for reachability analysis of discrete-time Markov processes. Redundant computations can be avoided by considering transitions with Dirac distributions. If two states of a Markov chain are connected with a Dirac transition, the iterative method can postpone the update of the source state until the value of the destination state is computed. Using this approach, we propose two heuristics to improve the performance of iterative methods for computing unbounded reachability probabilities in discrete-time Markov chains and Markov decision processes. The proposed heuristics can be lifted to the computations of expected rewards and have been implemented in PRISM model checker. Several standard case studies have been used and experimental results show that our heuristics outperform most well-known previous iterative methods.

Keywords: Probabilistic model checking · Quantitative verification · Markov Decision Processes · Discrete-time Markov chains

1 Introduction

Model checking is a formal approach for verifying quantitative and qualitative properties of computer systems. In this way, the system is modelled by a labelled transition system, and its properties are specified in temporal logic. Because of some stochastic behaviours of these systems, we can use probabilistic model checking to analyse the quantitative property specifications of these systems [1–3]. In this domain, we can use discrete and continuous time Markov Chains to model fully probabilistic systems. Besides, Markov Decision Processes [5] are used to model systems with both probabilistic and non-deterministic behaviours. Probabilistic Computation Tree Logic (PCTL) [1] is used to formally specify the

© IFIP International Federation for Information Processing 2020
Published by Springer Nature Switzerland AG 2020
L. S. Barbosa and M. Ali Abam (Eds.): TTCS 2020, LNCS 12281, pp. 1–16, 2020.
https://doi.org/10.1007/978-3-030-57852-7_1

related system properties. A main part of PCTL properties against MDPs can be verified by computing the extremal reachability probability: The maximal or minimal probability of reaching a given set of goal states. For quantitative parts, numerical computations are needed to calculate these reachability probabilities [2,6]. Linear programming [1,3], value iteration and policy iteration are well-known numerical approaches for computing the optimal reachability probabilities [2,5]. PRISM [4] and STORM [7] are examples of probabilistic model checkers that use these numerical methods to compute reachability probabilities.

One of the main challenges of model checking in all variants is the state space explosion problem, i.e., the excessive space requirement to store the states of the model in the memory [1,9]. For probabilistic model checking, we have the additional difficulty of solving linear programs. For the feasibility of the algorithms we need efficient heuristics to decrease the running time of these algorithms [3]. A wide range of approaches has been proposed for probabilistic model checking in previous works to tackle these problems. Symbolic model checking [9], compositional verification [10], symmetry reduction for probabilistic systems [11], incremental model construction [12] and statistical model checking [6] have been proposed to reduce the needed space for probabilistic model checking. In addition, several approaches are used to accelerate standard algorithms for probabilistic model checking. SCC-based approaches [13,14] identify strongly connected components (SCCs) of the underlying model and compute reachability probabilities of the states of each component in a right order. Learning based algorithms use the idea of real-time dynamic programming to solve reachability probability problems of MDPs [15]. Prioritization methods focus on finding a good state ordering to update the values of states during iterative computations [13,14]. The idea of finding Maximal End Components (MECs) is used in [3] to reduce the number of states of the model. Several techniques are proposed in [17] to reduce the size of a DTMC model. These techniques are used to reduce the model for finite-horizon properties.

In this paper, we propose new heuristics to reduce the number of updates and the running time of iterative methods for computing reachability probabilities. Our methods consider the set of transitions that lead to a Dirac distribution (a transition with probability one) and use the fact that any two states of a DTMC that are connected with a Dirac transition have the same (unbounded) reachability probabilities and the difference between their expected rewards equals to the defined reward of the transition. Although our heuristic can be considered as a special case of weak bisimulation, its time complexity is linear in the size of the underlying model. Our experiments show that an iterative method with our heuristic outperforms the one with bisimulation minimization. The main contributions of our work are as follows:

- We first propose a Dirac-based method to avoid redundant updates in the computation of reachability probabilities in DTMCs.
- To use our method for MDPs, we propose it as an extension of modified policy iteration method to compute the extremal reachability probabilities and expected rewards. We also apply the heuristic for SCC-based methods to use the benefits of SCC decomposition approaches.

A main advantage of our heuristic is its time complexity that is linear in the size of model. The remainder of this paper is structured as follows: Sect. 2 reviews some related definitions. In Sect. 3 we review the probabilistic model checking algorithms. Section 4 proposes the Dirac-based method to improve the iterative computations for reachability probabilities. Section 5 shows the experimental results and Sect. 6 concludes the paper.

2   Preliminaries

In this section, we provide an overview of DTMCs and MDPs and reachability properties. We mainly follow the notations of [1,13]. Let S be a countable set. A discrete probability distribution on S is a function $P : S \rightarrow [0,1]$ satisfying $\sum_{s \in S} P(s) = 1$. We use $Dist(S)$ as the set of all distributions on S. The support of P is defined as the set $Supp(P) = \{s \in S | P(s) > 0\}$. A distribution P is $Dirac$ if $Supp(P)$ has only one member. More details about the proposed definitions in this section and their related probability measures are available in [1,2,16].

2.1   Discrete-Time Markov Chains

Definition 1. A Discrete-time Markov Chain (DTMC) is a tuple $D = (S, \hat{s}, \mathbf{P}, R, G)$ where S is a countable, non-empty set of *states*, $\hat{s} \in S$ is the *initial state*, $\mathbf{P} : S \times S \rightarrow [0,1]$ is the *probabilistic transition function*, $R : S \times S \rightarrow \Re_{\geq 0}$ is a reward function wich assigns to each transition of P a non-negative reward value and $G \subseteq S$ is the set of *Goal* states.

A DTMC D is called finite if S is finite. For a finite D, $size(D)$ is the number of states of D plus the number of transitions of the form $(s, s') \in S \times S$ with $\mathbf{P}(s, s') > 0$. A *path* represents a possible execution of D [2] and is a non-empty (finite or infinite) sequence of states $s_0 s_1 s_2...$ such that $\mathbf{P}(s_i, s_{i+1}) > 0$ for all $i \geq 0$. We use $Paths_{D,s}$ to denote the set of all paths of D that start in the state s and we use $FPaths_{D,s}$ for the subset of *finite* paths of $Paths_{D,s}$. We also use $Paths_D$ and $FPaths_D$ for $\cup_{s \in S} Paths_{D,s}$ and $\cup_{s \in S} FPaths_{D,s}$ respectively. For a finite path $\pi = s_0 s_1...s_k$, the accumulated reward is defined as: $\sum_{i<k} R(s_i, s_{i+1})$. For an infinite path $\pi = s_0 s_1...$ and for every $j \geq 0$, let $\pi[j] = s_j$ denote the $(j+1)$th state of π and $\pi[..j]$ the $(j+1)$th prefix of the form $s_0 s_1...s_j$ of π. We use $pref(\pi)$ as the set of all prefixes of π.

2.2   Probability Measure of a Markov Chain

In order to reason about the behaviour of a Markov chain D, we need to formally use the *cylinder sets* of the finite paths of D [1].

Definition 2. The *Cylinder set* of a finite path $\hat{\pi} \in FPaths_D$ is defined as $Cyl(\hat{\pi}) = \{\pi \in Paths_D | \hat{\pi} \in pref(\pi)\}$. The probability measure Pr^D is defined on the cylinder sets as $Pr^D(Cyl(s_0...s_n)) = \prod_{0 \leq i < n} \mathbf{P}(s_i, s_{i+1})$.

2.3   Markov Decision Processes

Markov Decision Processes (MDPs) are a generalization of DTMCs that are used to model systems that have a combination of probabilistic and non-deterministic behaviour. An MDP is a tuple $M = (S, \hat{s}, Act, \delta, R, G)$ where S, $\hat{s}$ and G are the same as for DTMCs, Act is a finite set of actions, $R : S \times Act \times S \to \Re_{\geq 0}$ is a reward function, assigns to each transition a non-negative reward value and $\delta : S \times Act \to Dist(S)$ is a probabilistic transition function. For every state $s \in S$ of an MDP M one or more actions of Act are defined as enabled actions. We use $Act(s)$ for this set and define it as $Act(s) = \{\alpha \in Act \mid \delta(s, \alpha) \text{ is defined}\}$.

For $s \in S$ and $\alpha \in Act(s)$ we use $Post(s, \alpha)$ for the set of α successors of s, $Post(s)$ for all successors of s and $Pre(s)$ for predecessors of s [1]:

$$Post(s, \alpha) \doteq \{s' \in S \mid \delta(s, \alpha, s') > 0\}, \tag{1}$$

$$Post(s) \doteq \cup_{\alpha \in Act(s)} Post(s, \alpha), \tag{2}$$

To evaluate the operational behaviour of an MDP M we should consider two steps to take a transition from a state $s \in S$. First, one enabled action $\alpha \in Act(s)$ is chosen non-deterministically. Second, according to the probability distribution $\delta(s, \alpha)$, a successor state $s' \in Post(s, \alpha)$ is selected randomly. In this case, $\delta(s, \alpha)(s')$ determines the probability of a transition from s to s' by the action $\alpha \in Act(s)$. We extend the definition of *paths* for MDPs: A path in an MDP M is a non-empty (finite or infinite) sequence $\pi = s_0 \xrightarrow{\alpha_0} s_1 \xrightarrow{\alpha_1} \dots$ where $s_i \in S$ and $\alpha_i \in Act(s_i)$ and $s_{i+1} \in Post(s_i, \alpha_i)$ for every $i \geq 0$. Similar to the case with DTMCs, for a state $s \in S$, we use $Paths_{M,s}$ to denote the set of all paths of M starting in s and $FPaths_{M,s}$ for all finite paths of it. For reasoning about the probabilistic behaviour of an MDP we use the notion of policy (also called adversary) [2,3,6].

Definition 3. A (deterministic) *policy* of an MDP M is a function $\sigma : FPath_M \to Act$ that for every finite path $\pi = s_0 \xrightarrow{\alpha_0} s_1 \xrightarrow{\alpha_1} \dots \xrightarrow{\alpha_{i-1}} s_i$ selects an enabled action $\alpha_i \in Act(s_i)$. The policy σ is memoryless if it depends only on the last state of the path. In general, a policy is defined as function of $FPath_M$ to a distribution on Act. However, memoryless and deterministic policies are enough for computing the optimal unbounded reachability probabilities [2]. We use Pol_M for the set of all deterministic and memoryless policies of M. A policy $\sigma \in Pol_M$ resolves all non-deterministic choices in M and induces a DTMC M^σ for which every state is a finite path of M.

Definition 4 Induced DTMC. For an MDP $M = (S, \hat{s}, Act, \delta, R, G)$ and a policy σ, the induced DTMC is $M^\sigma = (FPath_M, \hat{s}, P^\sigma, G')$ where:

- Every path $\pi \in FPath_M$ is a state of M^σ
- For every $\pi = s_0 \xrightarrow{\alpha_0} s_1 \xrightarrow{\alpha_1} \dots \xrightarrow{\alpha_{n-1}} s_n$, $P^\sigma(\pi, \pi \xrightarrow{\alpha(s_n)} s_{n+1}) = \delta(s_n, \sigma(s_n))$ (s_{n+1}).
- $G' = \{s_0 s_1 \dots s_n \in FPath_M \mid s_n \in G\}$

2.4   Probability Measure for MDPs

For a policy σ and any infinite path $s_0 \xrightarrow{\alpha_0} s_1 \xrightarrow{\alpha_1} ... \in Paths_M$, we define the following bijection function f to associate the infinite paths in M and M^σ [2]:

$$f(s_0 \xrightarrow{\alpha_0} s_1 \xrightarrow{\alpha_1} ...) \overset{def}{=} (s_0)(s_0\alpha_0 s_1)...$$

where $\alpha_i = \sigma(s_i)$ for all $i \geq 0$. We can use this function and the probability measure Pr^{M^σ} for the induced DTMC M^σ to define the probability measure Pr_σ^M over $Paths_M$ [1,3]. We use Pr_σ^M to capture the behaviour of M under σ.

Although the number of states of the induced DTMC D is (countably) infinite, for memoryless policies its state space is isomorphic to S and we can reduce M^σ to an $|S|$-state *quotient DTMC*. The quotient DTMC for an MDP $M = (S, \hat{s}, Act, \delta, R, G)$ and a deterministic finite-memory policy σ is the finite state DTMC $M^\sigma = (S, \hat{s}, \mathbf{P}, R', G)$ where S, G and $\hat{s}$ are the same as in M, and $\mathbf{P} : S \times S \to [0, 1]$ is a transition probability function defined as $\mathbf{P}(s, s') = \delta(s, \sigma(s))(s')$ and R' is a reward function where $R'(s, s') = R(s, \sigma(s))(s')$ [2].

3   Probabilistic Model Checking

The aim of probabilistic model checking is to verify a desired quantitative or qualitative property of the system. A main class of PCTL properties includes reachability probabilities and expected rewards. For DTMCs, a reachability probability is the probability of reaching the set of goal states G and for MDPs, it is the extremal probability of reaching G. Reward-based properties are defined as the expected accumulated reward (extremal expected reward in the case of MDPs) until reaching a goal state [1]. To formally reason about reachability probabilities, we need to define a probability measure on the set of paths that reach to some states in G. For a state $s_0 \in S$, let $reach_{s_0}(G)$ be the set of all paths that start from s_0 and have a state from G:

$$reach_{s_0}(G) \overset{def}{=} \{\pi \in Paths_{D,s_0} \mid \pi[i] \in G \text{ for some } i \geq 0\}.$$

The probability measure on the set $reach_{s_0}(G)$ is defined as [1]:

$$Pr^D(reach_{s_0}(G)) = \sum_{s_0 s_1 ... s_n \in reach_{s_0}(G)} Pr^D(Cyl(s_0 s_1 ... s_n)).$$

For MDPs, reachability probabilities are defined as the extremal probabilities of reaching goal states [13, 16]:

$$Pr_{max}^M(reach_{s_0}(G)) \overset{def}{=} sup_{\sigma \in Pol_M} Pr_\sigma^M(reach_{s_0}(G))$$

Reachability probability properties are divided into qualitative and quantitative properties. A qualitative property of a probabilistic system means the probability of reaching the set of goal states is either 0 or 1 [1]. Qualitative verification is a method to find the set of states for which this reachability probability is exactly 0 or 1. We denote these sets by S^0 and S^1, respectively. For the case of MDPs, we are interested to find those states for which the maximum or minimum reachability probability is 0 or 1. These sets are denoted by S^0_{min}, S^0_{max}, S^1_{min} and S^1_{max} and can be computed by graph-based methods [2].

3.1 Quantitative Properties

In probabilistic model checking, we consider a PCTL property to be quantitative if the probability of reaching goal states is not exactly 0 or 1. Verification of quantitative properties usually reduces to solving a linear time equation system (for DTMCs) or solving a *Bellman* equation for MDPs [3,5]. For an arbitrary state $s \in S$ in a DTMC D, let x_s be the probability of reaching G from s, i.e., $x_s = Pr^D(reach_s(G))$. The values of x_s for all $s \in S$ are obtained as the unique solution of the *linear equation system* [1,2]:

$$x_s = \begin{cases} 0 & \text{if} \quad s \in S^0 \\ 1 & \text{if} \quad s \in S^1 \\ \sum_{s' \in S} \mathbf{P}(s, s').x_{s'} & \text{if} \quad s \in S^? \end{cases}$$

where $S^? = S \backslash (S^0 \cup S^1)$. A model checker can use any direct method (e.g. Gaussian elimination) or iterative method (e.g. Jacobi, Gauss-Seidel) to compute the solution of this linear equation system.

For the case of MDPs, we consider x_s as the maximum (or minimum) probability of reaching G, i.e., $x_s = Pr^M_{max}(reach_s(G))$. In this case, the values of x_s are obtained as the solution of the *Bellman equation* system:

$$x_s = \begin{cases} 0 & \text{if} \quad s \in S^0_{max} \\ 1 & \text{if} \quad s \in S^1_{max} \\ \max_{\alpha \in Act(s)} \sum_{s' \in S} \delta(s, \alpha)(s').x_{s'} & \text{if} \quad s \in S^?_{max} \end{cases}$$

where $S^?_{max} = S \backslash (S^0_{max} \cup S^1_{max})$. Using x_s for the maximal expected accumulated reward, we have [1,2]:

$$x_s = \begin{cases} 0 & \text{if } s \in G \\ \infty & \text{if } s \notin S^1_{min} \\ \max_{\alpha \in Act(s)} \sum_{s' \in S} (R(s, \alpha, s') + \delta(s, \alpha)(s').x_{s'}) & \text{otherwise} \end{cases}$$

Some standard direct algorithms (like Simplex algorithm [11]) are able to compute the precise values for the *Bellman equation*. The main drawback of direct methods is their scalability that limits them to relatively small models [2]. Iterative methods are other alternatives to approximate the values of x_s.

3.2 Iterative Methods for Quantitative Reachability Probabilities

Value iteration (VI) and policy iteration (PI) are two standard iterative methods that are used to compute the quantitative properties in probabilistic systems. VI and its Gauss-Seidel extension are widely studied in the previous works [1,2, 13,15,16]. We review PI, which is used in the remaining of this paper.

Policy Iteration. This method iterates over policies in order to find the optimal policy that maximizes reachability probabilities of all states. Starting from an arbitrary policy, it improves policies until reaching no change in them [2]. For each policy, the method uses an iterative method to compute the reachability probability values of quotient DTMCs and updates the value of states of the original MDP. The termination of this method is guaranteed for every finite MDP [2]. Modified policy iteration (MPI) [5] performs a limited number of iterations for each quotient DTMC (100 iterations for example) and updates the policy after doing this number of iterations. Algorithm 1 shows the standard policy iteration to compute $Pr_{max}^M(reach_s(G))$. We consider a policy as a mapping σ from states to actions. In lines 7–9, the algorithm uses a greedy approach to update the policy σ. More details about this algorithm are available in [2].

Algorithm 1. Policy iteration for $P_s^{max}(G)$.

---

input: an MDP $M = (S, \hat{s}, Act, \delta, G)$
output: Approximation of $Pr_{max}^M(reach_s(G))$ for all $s \in S$
for all $s \in S$ **do**

$$x_s \leftarrow \begin{cases} 1, & \text{if } s \in S_{max}^1 \\ 0, & \text{otherwise} \end{cases}$$

end for
select an arbitrary policy σ;
repeat
    Compute $x_s = Pr_\sigma^M(reach_s(G))$ for all $s \in S$;
    for all $s \in S^?$ **do**
        $\sigma(s) \leftarrow \text{argmax}_{a \in Act(s)} \sum_{s' \in S} \delta(s, a)(s') \times p_{s'}$;
    end for
until σ has not changed;

---

4 Improving Iterative Methods for Quantitative Reachability Probabilities

We use a heuristic to improve the performance of iterative methods for DTMCs. It considers transitions with Dirac distributions. For this class of transitions, the reachability probability values of the source and destination states of the transition are the same. The idea of our heuristic is to avoid useless updates in order to reduce the total number of updates before termination. We use it to improve the performance of policy iteration method for computing reachability probabilities in MDPs. In this case, Dirac transitions are considered to reduce the number of updates in the computations of each quotient DTMCs. We apply this technique on the SCC-based method to improve the performance of this approach. Although the idea of considering Dirac transitions has been used as a reduction technique in [8], it has been only proposed for the case of deterministic Dirac transitions. The main contribution of our work is to use this heuristic for improving the MPI method for both reachability probabilities and expected rewards to cover nondeterministic action selections. To the bet of our knowledge, none of state of the art model checkers support this technique.

4.1   Using Transitions with Dirac Probability

In our work, we consider transitions with Dirac distribution (which we call Dirac transitions) to avoid redundant updates of state values. This type of transitions is also used in statistical model checking [6] but we use it to accelerate iterative methods. For a DTMC D, a Dirac transition is a pair of states s and s' with $\mathbf{P}(s, s') = 1$. Based on the definition, for this pair of states, we have $Pr^D(reach_s(G)) = Pr^D(reach_{s'}(G))$. As a result, we can ignore this transition and postpone the update of the value of s until the convergence of the value of s'. Our approach updates the value of s only once and avoids redundant updates. However, there may be some incoming transitions to s that need the value of s in every iteration. Consider Fig. 1 and suppose s is the target state of a transition of the form (t, s). In this case the value of s affects the value of t and we should modify the incoming transitions to s to point to s'. Algorithm 2 shows details of our approach to remove Dirac transitions from a DTMC D and reduce it to smaller DTMC D'. The main idea of this algorithm is to partition the state space of D to *Dirac-based* classes. A *Dirac-based* class is the set of states that are connected by Dirac transitions. The algorithm uses the DBC array to partition $S^?$ to the related classes. Consider for example a sequence of states of the form $s_i, s_j, ..., s_{k-1}, s_k$ where there exists Dirac transitions between each two consecutive states and the outgoing transition of s_k is not a Dirac one or s_k is one of $s_i, ..., s_{k-1}$. We put these states into one class and set their DBC to s_k. The reachability probability of all states of this class are equal to the reachability probability of s_k. Formally, for each state $s_i \in S$, we define $DBC[s_i]$ as the index of the last state in the sequence of states that are connected with *Dirac* transitions. For every state $s \in S$ we have $x_s = x_{DBC[s]}$. Algorithm 2 initiates the values of this array in lines 3–5. For every state $s_i \in S$ that $DBC[s_i]$ is

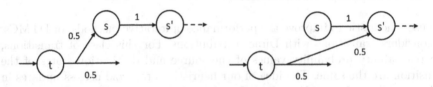

Fig. 1. An example of a Dirac transition (left) and the modified DTMC (right)

not determined previously, the algorithm calls the $Find\_Dirac\_Index$ function to determine its DBC. After determining the DBC value of all states of S, the algorithm creates the reduced DTMC D' (lines 9–13). The states of D' are those states $s_i \in S$ for which $DBC[s_i] = i$. According to the definition of S^1, for each state $s \in G$ and each $s' \in S$ if $DBC[s'] = DBC[s]$ then we have $s' \in G$.

Algorithm 2. Dirac transition reduction in a DTMC D

1: **input:** a DTMC $D = (S, \hat{s}, \mathbf{P}, G)$
2: **output:** $P_s^{max}(F)$ for all $s \in S^?$
3: **for all** $s_i \in S^?$ **do**
4:    $DBC[s_i] = -1$;
5: **end for**
6: **for all** $s_i \in S^?$ **do**
7:    **if** $DBC[s_i] = -1$ **then**
8:       $DBC[s_i] = Find\_Dirac\_Index(D, DBC, s_i)$;
9:    **end if**
10: **end for**
11: Create reduced DTMC $D' = \{S', \hat{s}, \mathbf{P}', G\}$ where:
12: $S' = \{s_i \in S | DBC[s_i] = i\}$;
13: **for all** $(s, s') \in S'$ **do**
14:    $\mathbf{P}'(s, s') = \sum_{\{s'' \in S | DBC[s''] = DBC[s']\}} \mathbf{P}(s, s'')$
15: **end for**
16: **return** D';

Algorithm 3 shows details of the $Find\_Dirac\_Index$ function to determine the DBC value of a state s_i. It first checks the number of outgoing transitions of s_i. If there is more than one outgoing transition, the state should be considered in the reduced DTMC D' and its DBC value is set to the index of the state (line 11). If there is only one outgoing transition (a Dirac one of the form (s_i, s_j)), the

---

Algorithm 3. Computing $Find\_Dirac\_Index(s_i)$

1: **input:** a DTMC $D = (S, \hat{s}, \mathbf{P}, G)$, a DBC array and a state $s_i \in S$
2: **output:** $DBC[s_i]$;
3: **if** there is only one state s_j in $Post(s_i)$ and s_j is unmarked **then**
4:    **if** $DBC[s_j] = -1$ **then**
5:       Mark s_j;
6:       $DBC[s_i] = Find\_Dirac\_Index(D, DBC, s_j)$;
7:    **else**
8:       $DBC[s_i] = DBC[s_j]$;
9:    **end if**
10: **else**
11:    $DBC[s_i] = i$;
12: **end if**
13: **return** $DBC[s_i]$;

---

algorithm checks the DBC value of $s_j \in Post(s_i)$ and recursively computes the DBC values of the sequence of states that are connected via Dirac transitions.

The following lemma shows that the reachability probabilities of the states of D' are the same as the values of the corresponding states of D. As a result, the iterative method can be applied on D' which may be smaller than D.

Lemma 1: For each state $s_i \in S^?$ with $DBC[s_i] = i$ we have:
$Pr^{D'}(reach_{s_i}(G)) = Pr^D(reach_{s_i}(G))$ where D' is the Dirac-based reduced DTMC.

Proof: Recall that for each $s_i \in S^?$ we have

$$x_{s_i} = Pr^D(reach_{s_i}(G)) = \sum_{s' \in Post(s_i)} P(s_i, s').x_{s'} \tag{3}$$

Suppose $Post(s_i) = \{s'_1, s'_2, ..., s'_k\}$. First consider a case where for each two states $s'_i, s'_j \in Post(s_i)$ we have $DBC[s_i] \neq DBC[s_j]$ and as a result, for each $s'_j \in Post(s_i)$ we have $P'(s_i, DBC[s'_j]) = P(s_i, s'_j)$. We can rewrite equation (4) as:

$$x_{s_i} = P(s_i, s'_1).x_{s'_1} + ... + P(s_i, s'_k).x_{s'_k} = P(s_i, s'_1).x_{DBC[s'_1]} + ... + P(s_i, s'_k).x_{DBC[s'_k]}$$

$$= P'(s_i, DBC[s'_1]).x_{DBC[s'_1]} + ... + P'(s_i, DBC[s'_k]).x_{DBC[s'_k]} = Pr^{D'}(reach_{s_i}(G))$$

Now consider a case that $Post(s_i) = \{s'_{11}, s'_{12}, ..., s'_{1k}, ..., s'_{l1}, ..., s'_{lm}\}$ where the indexes are so that for each s'_{ij} and s'_{pq} we have $DBC[s'_{ij}] = DBC[s'_{pq}]$ if $i = p$ and $DBC[s'_{ij}] \neq DBC[s'_{pq}]$ if $i \neq p$. Using the definition of P' we have:

$$x_{s_i} = P(s_i, s'_{11}).x_{s'_{11}} + ... + P(s_i, s'_{1k}).x_{s'_{1k}} + ...$$

$$+ P(s_i, s'_{l1}).x_{s'_{l1}} + ... + P(s_i, s'_{lm}).x_{s'_{lm}}$$

$$= P(s_i, s'_{11}).x_{DBC[s'_{11}]} + ... + P(s_i, s'_{1k}).x_{DBC[s'_{1k}]} + ...$$

$$+ P(s_i, s'_{l1}).x_{DBC[s'_{l1}]} + ... + P(s_i, s'_{lk}).x_{DBC[s'_{lk}]}$$

$$= [P(s_i, s'_{11}) + ... + P(s_i, s'_{1k})].x_{DBC[s'_{11}]} + ...$$

$$+ [P(s_i, s'_{l1}) + ... + P(s_i, s'_{lk})].x_{DBC[s'_{l1}]}$$

$$= P'(s_i, DBC[s'_{11}]).x_{DBC[s'_{11}]} + ... + P'(s_i, DBC[s'_{l1}]).x_{DBC[s'_{l1}]} = Pr^{D'}(reach_{s_i}(G)). \quad \blacksquare$$

The time complexity of Algorithm 2 is linear in the size of the DTMC D. For every state $s \in S^?$, the *Find_Dirac_Index* function is called once and performs a check on the number of outgoing transitions of s and its DBC value. For creating the reduced DTMC D', Algorithm 2 checks every state and transitions of D only once. As a result the time complexity of Algorithm 2 is in $O(size(D))$.

The iterative method (Gauss-Seidel for example) can then compute the probability values of the states of D'. Algorithm 4 shows the way that the Dirac-based

Algorithm 4. Dirac-Based Reduction Technique for Computing the Reachability Probabilities of DTMCs

---

1: **input:** a DTMC $D = (S, \hat{s}, \mathbf{P}, G)$
2: **output:** Reachability probabilities of the states of D;
3: Apply Algorithm 2 on D and compute the reduced DTMC D';
4: Apply Gauss-Seidel method for approximating reachability probabilities of S' in D';
5: **for all** $s \in S^?$ **do**
6:    $x_s = x_{DBC[s]}$;
7: **end for**
8: return $(x_s)_{s \in S}$;

DTMC-reduction technique is used to accelerate the computations of reachability probabilities.

It first computes the reduced DTMC D' and then computes reachability probabilities of states of D'. Finally, in lines 3–5, it updates the value of the remaining states of D (those states that are not in D'). For the correctness of Algorithm 4, we compare the precision of the computed values of this algorithm with the computed values of Gauss-Seidel algorithm for DTMCs. We suppose that both algorithms use the same state ordering, i.e., for each i, j they update s_i before s_j if $i < j$. In this case we have

$$x_{s_i}^k = \sum_{s_j \in Post(s_i), j < i} \mathbf{P}(s_i, s_j) \times x_{s_j}^k + \sum_{s_j \in Post(s_i), j > i} \mathbf{P}(s_i, s_j) \times x_{s_j}^{k-1} \quad (4)$$

Lemma 2: Consider a DTMC D, its Dirac-based reduced DTMC D' and a state $s_i \in S$. Let $x_{s_i}^k$ be the computed reachability probability of s_i after k iteration of Gauss-Seidel method on D and $x'^k_{s_i}$ be the computed value of $s_{DBC[s_i]}$ after k iteration of Gauss-Seidel method on D'. For every $k \geq 0$ we have $x'^k_{s_i} \geq x_{s_i}^k$.
Proof Sketch: By induction on k and the fact that for each state $s_j \in S$ we have $x_{s_j}^k \leq x_{DBC[s_j]}^k$. ∎

According to Lemma 2, the precision of the computed values of Algorithm 4 is the same or better than the precision of the computed values of standard Gauss-Seidel if we use the same number of iterations and the same state ordering in both methods. Depending to the structure of the DTMC D, either of two algorithms may converge faster than the other algorithm. Faster convergence may cause non-precise results for some values. In general, standard iterative methods does not guarantee the convergence of values [16].

4.2 Improving Iterative Methods for Computing Reachability Probabilities in MDPs

The Dirac-based reduction technique can be also used for MDPs to avoid redundant updates. However, it can not be directly applied for states with multiple enabled actions. To cover this case and as a novelty of our approach, we use the MPI method. We apply our heuristic for every quotient DTMC to reduce the number of states that should be updated. We propose our it as an extension of SCC-based techniques. We use MPI to compute reachability values of the states of each SCC and apply our Dirac-based heuristic to accelerate the iterative computations of each quotient DTMC. Algorithm 5 presents the overall idea of our method for accelerating SCC-based methods for MDPs. The correctness of this approach relies on the correctness of SCC decomposition for MDPs [13] and the correctness of our Dirac-based DTMC reduction method (Lemma 1).

Algorithm 5. SCC-based method with Dirac-based reductions for computing reachability probabilities.

1: **input:** an MDP $M = (S, \hat{s}, Act, \delta, G)$
2: **output:** $Pr^M_{max}(reach_s(G))$ for all $s \in S$;
3: Compute Strongly Connected Components of M and order them by the reverse topological order $C_1, ..., C_k$
4: **for** i = 1 to k do **do**
5:    Select a policy σ of C_i
6:    **repeat**
7:        Use σ and compute the quotient DTMC D_i of C_i ;
8:        Compute Dirac-based reduced DTMC D'_i;
9:        Use an iterative method to update the value of states of D'_i.
10:       Update the remaining states of D_i (like lines 3-5 of algorithm 5).
11:       Update σ according to a greedy approach.
12:    **until** σ has changed;
13: **end for**
14: return $(x_{s_i})_{s_i \in S?}$;

4.3 Accelerating Iterative Methods for Computing Expected Rewards in MDPs

The standard iterative methods can be used to approximate the Bellman equation for extremal expected rewards. In this case, the initial vector of values is set to zero for all states. Value iteration (or policy iteration) should also consider the defined reward of each action in the update of values of each state. More details about iterative methods for expected rewards are available in [1,2,16]. To use our heuristic for expected rewards, we use the fact that for every Dirac transition of the form (s_i, s_j) we have $x_{s_i} = x_{s_j} + R(x_i)$ where x_i and x_j are the expected values for these two states and $R(x_i)$ is the reward for x_i. In this case, an iterative method does not need to update the value of x_i in every iteration. Similar to the case for the reachability probabilities, the incoming transitions to x_i should be modified to point to x_j. In addition the reward of a modified transition is modified by adding the reward of the related Dirac transition.

5 Implementation and Experimental Results

We implemented our heuristics as a package in the PRISM model checker. Our implementations are based on sparse engine of PRISM [9] which is developed in C and are available in [18]. We used several standard case studies which have been used in previous works [2,4,10,13,15,16]. We compare the running time of our heuristics with the running time of well-known previous methods. We only focus on the running time of the iterative computations for quantitative properties. We exclude the running times for model constructions that are the same for all methods. The running time of the graph-based pre-computation are negligible in appropriate implementations [7]. For SCC-based methods and our Dirac-based ones, the reported times include the running times for SCC decomposition and Dirac-based reductions. All benchmarks have been run on a machine with Corei7 CPU (2.8 GHz, 4 main cores) and 8 GB RAM running Ubuntu 18. We use *Consensus, Zeroconf, firewire_abstract*, brp, *nand* and *Crowds* case studies for comparing the performance of iterative methods for reachability probabilities and use *Wlan, CSMA, Israeli-jalefon* and *Leader* cases for expected rewards. These case studies are explained in [4,15,16]. More details about our experiments and model parameters are available as log-files in [18]. Although there are several other standard case studies, their graphical structure do not have any cycle and a topological backward iterative method can be used to computed their underlying properties in linear time [13,14]. Hence, we focus on the case studies of Table 1, where have several non-trivial cycles.

We consider the running time of the standard iterative methods and well known improved techniques from previous works. To perform a fair comparison, we use sparse engine of PRISM for all experiments and we also implemented topological (SCC-based) method for this engine. In this case, we use MPI to solve each SCC. For learning-based methods, we use the implementation that is proposed in [15] and for backward value iteration, we implement the proposed method from [14]. For all case studies, we consider $\epsilon = 10^{-8}$ as the threshold.

Table 1 shows the running time of the iterative methods for MDP models. For the SCC-based method, we use MPI to approximate the reachability probability values of each SCCs. All times are in seconds. We use * where a method does not terminate after an hour. For *consensus, Israeli-jalefon, Leader* and *Wlan* models, the running time of SCC-based method is less than the others. In these cases, we use our Dirac-based method to reduce the running time of the computations for each SCC. The results show that for these two classes, our technique is faster than the other methods. The learning-based method is faster than other methods for *zeroconf* cases with N = 20. For *firewire* case studies, SCC based and backward value iteration methods are much faster than the standard iterative methods. In this case Dirac-based method for MPI (without SCC decomposition) works better than the other methods.

Table 2 shows the results of our experiments for DTMC case studies. We present the running time of Jacobi and Gauss-Seidel as the standard methods and SCC-based and backward Gauss-Seidel as the improved methods from the previous works. We also use the STORM model checker with bisimulation minimization technique to compare its impact on the running time of computations. The results of Table 1 and Table 2 show that our proposed method outperforms all previous standard methods and most improved ones.

Table 1. Running times of the iterative methods for quantitative properties of MDPs.

Model	GS-VI	MPI	Learning-based	Backward-VI	SCC-based GS	SCC-based MPI	Dirac-based
Consensus 4, 20	354.7	190.2	*	273	27.7	38.25	13.12
Consensus 4, 36	1812	968.6	*	1376	111.7	226.5	76.55
Consensus 5, 15	1966	441.3	*	1421	183.3	101.1	35.61
Consensus 5, 32	*	*	*	*	1327	2048	706
Consensus 6, 8	2680	1082	*	1783	430.3	281.6	101.2
Zeroconf 14 (N = 20)	16.77	41.16	2.23	14.45	12.67	51.91	5.2
Zeroconf 16 (N = 20)	20.11	41.84	2.3	16.11	13.8	71.54	6.92
Zeroconf 18 (N = 20)	22.6	52.63	2.47	17.7	14.12	81.89	8.07
Zeroconf 14 (N = 5000)	27.9	58.52	731.2	23.71	12.73	69.9	6.54
Zeroconf 16 (N = 5000)	32.74	62.70	854.3	25.9	13.85	79.0	6.83
Zeroconf 18 (N = 5000)	36.46	64.31	878.4	27.54	14.57	104.46	12.74
firewire_abst 7000,128	473.3	260.6	742.9	1.17	17.14	5.34	0.52
firewire_abst 9500,64	585.5	323.6	574.4	1.23	22.4	4.82	0.44
firewire_abst 9500,128	887.7	487.5	917.6	1.42	27.39	O.M	0.52
Wlan 5,1500	56.76	42.99	*	27.18	6.76	1.61	1.15
Wlan 5,3000	99.07	74.75	*	53.82	16.62	2.62	2.06
Wlan 6,250	160.1	121.5	*	97.5	12.17	3.65	1.44
Wlan 6,1200	257.3	195.9	*	182.4	22.9	4.82	2.67
CSMA 4,3	32.1	32.5	*	15.4	10.6	4.66	4.17
Israeli-jalefon-20	117	47.2	*	39.3	13.6	2.88	2.26
Israeli-jalefon-22	203	71.8	*	55.7	24.9	13.1	11.7
Leader 7	3.03	1.11	*	2.45	0.95	0.67	0.13
Leader 8	22.4	6.49	*	13.7	8.41	6.59	2.13

Table 2. Running times of computations for reachability probabilities of DTMCs.

Model	Jacobi	Gauss-Seidel	SCC-based	Backward	Dirac-based	Bisimulation
brp 4,8192	296	137	13.2	32.7	6.14	453
brp 6,4096	104	47.3	7.5	19	4.1	349
brp 6,8192	415	233	28.5	36.2	11.7	O.M
crowds 10, 10	32.07	11.67	21.8	7.48	2.46	5.58
crowds 12, 10	150.89	49.26	*	41.8	11.49	27.3
crowds 9, 15	61.53	19.66	37.14	12.55	5.00	O.M
nand 5, 60	620.7	554.9	8.7	23.72	143.25	244.6
nand 6, 50	364.8	326.2	4.83	11.22	83.60	246.1
nand 6, 60	885.3	797.2	*	38.56	204.1	247.2

6   Conclusion

In this paper, we proposed Dirac-based heuristics to accelerate iterative methods for computing reachability probabilities and expected rewards. These heuristics have been proposed to avoid redundant updates and multiplications of state values. They can be used for the Gauss-Seidel method for DTMCs or MPI for MDPs. Experiments show promising results. In most cases, our heuristics reduce the running time of iterative methods to less than 50% of the running time of the best previous methods. The extension of the proposed heuristics for other classes of properties (LTL properties for example) is an interesting direction for future works. The Dirac-based method reduces data dependency and can be used for parallel version of the Gauss-Seidel and MPI methods.

References

1. Baier, C., Katoen, J.P.: Principles of Model Checking. MIT Press, Cambridge (2008)
2. Forejt, V., Kwiatkowska, M., Norman, G., Parker, D.: Automated verification techniques for probabilistic systems. In: Bernardo, M., Issarny, V. (eds.) SFM 2011. LNCS, vol. 6659, pp. 53–113. Springer, Heidelberg (2011). https://doi.org/10.1007/978-3-642-21455-4_3
3. Baier, C., de Alfaro, L., Forejt, V., Kwiatkowska, M.: Probabilistic model checking, dependable software. Syst. Eng. **45**, 1–23 (2016)
4. Kwiatkowska, M., Norman, G., Parker, D.: The PRISM benchmark suite. In: 9th International Conference on Quantitative Evaluation of SysTems, pp. 203–204. IEEE CS Press (2012)
5. Puterman, M.L.: Markov Decision Processes: Discrete Stochastic Dynamic Programming. Wiley, Hoboken (2014)
6. Hartmanns, A.: On the analysis of stochastic timed systems. Ph.D. thesis, Saarland University (2015)
7. Dehnert, C., Junges, S., Katoen, J.-P., Volk, M.: A STORM is coming: a modern probabilistic model checker. In: Majumdar, R., Kunčak, V. (eds.) CAV 2017. LNCS, vol. 10427, pp. 592–600. Springer, Cham (2017). https://doi.org/10.1007/978-3-319-63390-9_31
8. D'Argenio, P.R., Jeannet, B., Jensen, H.E., Larsen, K.G.: Reduction and refinement strategies for probabilistic analysis. In: Hermanns, H., Segala, R. (eds.) PAPM-PROBMIV 2002. LNCS, vol. 2399, pp. 57–76. Springer, Heidelberg (2002). https://doi.org/10.1007/3-540-45605-8_5
9. Parker, D.A.: Implementation of symbolic model checking for probabilistic systems. Ph.D. thesis, University of Birmingham (2003)
10. Feng, L.: On learning assumptions for compositional verification of probabilistic systems. Ph.D. thesis, University of Oxford (2013)
11. Kwiatkowska, M., Norman, G., Parker, D.: Symmetry reduction for probabilistic model checking. In: Ball, T., Jones, R.B. (eds.) CAV 2006. LNCS, vol. 4144, pp. 234–248. Springer, Heidelberg (2006). https://doi.org/10.1007/11817963_23
12. Ujma, M.: On Verification and controller synthesis for probabilistic systems at runtime. Ph.D. thesis, University of Oxford (2015)

13. Kwiatkowska, M., Parker, D., Qu, H.: Incremental quantitative verification for Markov decision processes. In: IEEE/IFIP 41st International Conference on Dependable Systems & Networks (DSN), pp. 359–370. IEEE (2011)
14. Ciesinski, F., Baier, C., Groesser, M., Klein, J.: Reduction techniques for model checking Markov decision processes. In: Fifth International Conference on Quantitative Evaluation of Systems, pp. 45–54. IEEE (2008)
15. Brázdil, T., et al.: Verification of Markov decision processes using learning algorithms. In: Cassez, F., Raskin, J.-F. (eds.) ATVA 2014. LNCS, vol. 8837, pp. 98–114. Springer, Cham (2014). https://doi.org/10.1007/978-3-319-11936-6_8
16. Baier, C., Klein, J., Leuschner, L., Parker, D., Wunderlich, S.: Ensuring the reliability of your model checker: interval iteration for Markov decision processes. In: Majumdar, R., Kunčak, V. (eds.) CAV 2017. LNCS, vol. 10426, pp. 160–180. Springer, Cham (2017). https://doi.org/10.1007/978-3-319-63387-9_8
17. Kamaleson, N.: Model reduction techniques for probabilistic verification of Markov chains. Ph.D. thesis, University of Oxford (2018)
18. https://github.com/sadeghrk/prism/tree/DiracBased-Improving

Combining Machine and Automata Learning for Network Traffic Classification

Zeynab Sabahi-Kaviani[1], Fatemeh Ghassemi[1,2](✉), and Zahra Alimadadi[1]

[1] School of Electrical and Computer Engineering,
University of Tehran, Tehran, Iran
`fghassemi@ut.ac.ir`
[2] School of Computer Science, Institute for Research in Fundamental Sciences,
PO.Box 19395-5746, Tehran, Iran

Abstract. Viewing the generated packets of an application as the words of a language, automata learning can be used to derive the behavioral packet-based model of applications. The alphabets of the learned automata, manually defined in terms of packets, may cause overfitting. As some packets always appear together, we apply machine learning techniques to automatically define the alphabet set based on the timing and statistical features of packets. Using the learned automata models, the classifier should detect the accepted words of the models in the input. To improve this time-consuming process, we present a framework, called NeTLang, that identifies the application model in terms of k-testable languages. The classification problem is reduced to observing only $\Theta(k)$ symbols from the input with the help of machine learning techniques. Our framework utilizes the two diverse automata learning and machine learning techniques to build on their strengths (to be fast and accurate) and to eliminate their weaknesses (i.e., ignoring temporal relations among packets). According to our results, NeTLang outperforms the state-of-the-art methods using each technique alone.

Keywords: Automata learning · Machine learning · Traffic classification · Model inference

1 Introduction

Characterizing the network traffic and identifying running applications play an important role in several network administration tasks such as protecting against malicious behaviors, firewalling, and balancing bandwidth usage. Recently, dynamic port assignment and encryption protocol usage have considerably reduced the performance of the classic traffic classification methods, including port-based and deep packet inspection. This leads researchers to apply Machine-Learning techniques and behavioral pattern detection for traffic classification. Machine-Learning approaches classify network traffic based on statistical

© IFIP International Federation for Information Processing 2020
Published by Springer Nature Switzerland AG 2020
L. S. Barbosa and M. Ali Abam (Eds.): TTCS 2020, LNCS 12281, pp. 17–31, 2020.
https://doi.org/10.1007/978-3-030-57852-7_2

features with the granularity of flow [1,2] or packet [3], and hence, they ignore temporal relations among flows, and as a result, their false positive rates are not negligible although they are fast. In behavioral classification methods [4–6], an expert extracts specific behavioral aspects of a particular application or application type for the classification purpose. For instance, link establishment topology was used as the distinctive metric to classify P2P-TV traffic in [5].

Assuming a packet trace as a word of the language of an application, one can derive an automaton modeling the traffic behavior of that application. Automata learning approaches have been recently used to automatically derive the model of applications [7,8], network protocols [9,10], or Botnet behavior [8]. The alphabets of the learned automata are either manually defined by domain experts which is not straightforward, or in terms of packets which may cause overfitting. As some packets always appear together, we can consider a sequence of related packets together as a symbol of the alphabet. To this aim, we apply machine learning techniques to automatically define the alphabet set based on the timing and statistical features of packets.

The derived automata are used for traffic classification. A packet trace is classified into an application if the model of that application accepts it. Using automata learning methods, the classification problem is constrained to observe the complete trace of an application to verdict its acceptance/rejection. To tackle this challenge, inspired by [11], we upgrade the detection of an application based on partial observation of a trace, a window of size k, and derive a model that accepts a k-Testable language in the Strict Sense (k-TSS) [12]. K-TSS, a class of regular languages, also known as *window language*, allows to locally accept or reject a word by a sliding window parser of size k. We relax the acceptance condition of automata learning using machine learning by defining a proximity metric to be compatible with the local essence of the learned language. The proposed proximity metric is defined as a distance function. We have implemented our approach in a framework called Network Traffic Language learner, NeTLang. We evaluate the performance of our approach by applying it to real-world network traffic and compare it with machine and automata learning approaches. We achieved F1-Measure of 97% for both application identification and traffic characterization tasks. In summary, first, we learn the alphabet using a machine learning technique. Then, the network language is learned through an automata learning approach. Finally, the classifier identifies the classes based on our defined distance metrics on the input and the learned models. Our method makes these **contributions**: 1) Utilizing locally testable language learning in the traffic classification problem, 2) Extracting the domain-based alphabet automatically, 3) Upgrading the word acceptance by a new proximity metric. With these contributions, the following improvements are brought into traffic classification:

- Considering a sequence of related packets as the appropriate granularity of the problem, instead of per-packet detection which is too fine-grained or per-flow detection which is too coarse-grained,

- Providing highly accurate models for applications as our automata learning approach considers the temporal relation among flows and the way they are interleaved,
- Decreasing the classification time by considering only some first packets of a trace with a help of a novel distance function.

2 Background on Automata Learning

In this section, we provide some background on automata learning concepts used in our methodology. Learning a regular language from given positive samples (words belonging to the language) is a common problem in grammatical inference. To solve this problem, many algorithms were proposed to find the smallest deterministic finite automaton (DFA) that accepts the positive examples. In this paper, we focus on learning k-testable languages in the strict sense, a subset of a regular language, called k-TSS, initially was introduced by [12]. In such a language, words are determined by allowed three sets of prefixes and suffixes of length $k - 1$ and substrings of length k. It has been proven that it is possible to learn k-TSS languages in the limit [13]. To learn this language, the only effort is to scan the accepting words while simultaneously constructing the allowed three set. The locally testable feature of this language makes it appropriate for network traffic classification and other domains, such as pattern recognition [14] and DNA sequence analysis [15]. In the following, we provide the formal definition of k-TSS language taken from [16].

Definition 1 (k-test Vector). *Let $k > 0$, a k-test vector is determined by a 5-tuple $Z = \langle \Sigma, I, F, T, C \rangle$ where*

- *Σ is a finite alphabet,*
- *$I \subseteq \Sigma^{(k-1)}$ is a set of allowed prefixes of length less than k,*
- *$F \subseteq \Sigma^{(k-1)}$ is a set of allowed suffixes of length less than k,*
- *$T \subseteq \Sigma^k$ is a set of allowed segments, and*
- *$C \subseteq \Sigma^{<k}$ contains all strings of length less than k.*

Definition 2 (k-TSS Language). *Let $Z = \langle \Sigma, I, F, T, C \rangle$ be a k-test vector, for some $k > 0$. Then Language $\mathcal{L}(Z)$ in the strict sense (k-TSS) is computed as:*

$$\mathcal{L}(Z) = [(I\Sigma^* \cap \Sigma^* F) - \Sigma^*(\Sigma^k - T)\Sigma^*] \cup C$$

For instance, consider $k = 3$ and $Z = \langle \Sigma = \{a, b\}, I = \{ab\}, F = \{ab, ba\}, T = \{aba, abb, bba\}, C = \{ab\} \rangle$, then, $aba, abba \in \mathcal{L}(Z)$ since they are preserving the allowed sets of Z, while $bab, abb, abab, a$ do not belong to $\mathcal{L}(z)$ because, in order, they violate I ($ba \notin I$), F ($bb \notin F$), T ($bab \notin T$), and C ($a \notin C$). To construct the k-test vector of a language, we scan the accepted word by a k-size frame. By scanning the word $abba$ by a 3-size frame, ab, ba, and abb, bba are added to I, F, and T, respectively.

In our problem, we produce words such that their length is greater than or equal to k. It means that C always is empty. Hence, for simplicity, we eliminate C from the k-TSS vector for the rest of the paper.

3   Problem Statement and Basic Definitions

In this section, we first formally state the problem, and then we define some network concepts used in our methodology.

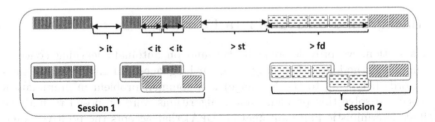

Fig. 1. Schematic representation of applying the timing parameters on a packet trace. Different backgrounds represent different network flows, and the floating boxes in the bottom represent derived network units.

Once a user runs an application, several *network connections* are established for transferring necessary data and control information, named network flow. A *network flow* consists of a sequence of packets, sent or received, having the same source IP, source port, destination IP, destination port, and protocol. Due to concurrency of the network, packet sequences of flows may be interleaved. In other words, the packet trace of an application is an interleaving of several flows.

We assume that for the applications set $A = \{App_1, App_2, ..., App_m\}$, the collection Π, consisting of the network traces, is given as an input, where $\pi_{App_i} \in \Pi$ is the captured traffic of App_i over a period of time. The goal is to distinguish which application(s) is running on a system by exploring the system traffic, named π. As the output, we label different parts of π with appropriate labels of application names.

To solve the problem, the main task is generating application models to be used by a classifier. Our intuition is that each application App_i has its network communicating language $\mathcal{L}(App_i)$ which differs from others. Since learning a language requires some of its positive examples, an important subproblem is to break down the input trace π_{App_i} of application App_i to a list of words $W_i = [w_1, w_2, ...w_n]$, where $w_j \in \mathcal{L}(App_i)$ is a successive subsequence of π_{App_i}. Instead of defining the alphabet in terms of packets, we consider a symbol for the sequence of packets that always appear together. To define the alphabet, we define some network concepts used in breaking down a trace:

Definition 3 (Session). *A session is a sequence of packets to/from the same user IP addresses where every two consecutive packets have a time interval of less than a **session threshold** (st). We have defined this concept to consider the latency when switching between two applications or within two different tasks of the execution of an application.*

Definition 4 (Flow-Session). *A Flow-Session is defined based on flows' inactive timeout. A flow is considered to be expired when it has been idle for a duration of more than a threshold called **inactive timeout** (it). Each flow in a session is split into shorter units called flow-session when two consecutive packets have a time interval of more than the inactive timeout. We have considered this concept to distinguish different phases of an application task.*

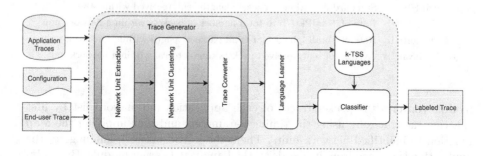

Fig. 2. The architectural view of NeTLang.

According to these definitions, for instance, a new *session* is started when switching from a Commit command in Git bash to an Add command, while each of these commands comprises some *flow-sessions* corresponding to different steps of executing them such as initializing or transferring data.

Definition 5 (Network Unit). *With the interleaved feature of network flows, to be able to observe the relation among flows, each flow-session is chunked into shorter units having the duration of at most a **flow duration** (fd), named network unit. The flow duration criterion is another timing constraint of our problem expressing an upper bound time until when a flow can last.*

Figure 1 shows an example of applying the above concepts on a trace. The trace is split into two *sessions* based on the *st*. The *flow-sessions* are extracted based on the *it* metric, and also, the *fd* threshold leads to the creation of smaller flow parts defined as network units. From the above definitions, by using the three configuration variables naming *st*, *fd*, and *it*, we break an input into some sub-traces to identify the network units.

We define function *Map: Network Units → Σ* to map the derived network units to the *network alphabet* to be processable by the automata techniques. Each symbol $s \in \Sigma$ stands for an unbreakable packet sequence of a flow.

4   Methodology

Figure 2 illustrates the overall scheme of our methodology. There are three main modules *Trace Generator*, *Language Learner*, and *Classifier*. The trace generator

Table 1. Flow statistical features: the f/b stands for forward/backward flow.

#	Feature	Description
1	TotalPktf/b	Count of packets sent/received within a network unit
2	TotalLf/b	Sum of packets' length sent/received within a network unit
3	MinLf/b	Minimum length of packet sent/received within a network unit
4	MeanLf/b	Average length of packet sent/received within a network unit
5	MaxLf/b	Maximum length of packet sent/received within a network unit
6	StdLf/b	Standard deviation of packets length sent/received within a unit
7	PktCntRf/b	Rate of TotalPktf/b to total number of packets within a network unit
8	DtSizeRf/b	Rate of TotalLf/b to sum of all packets' length within a network unit
9	AvgInvalf/b	Average of sent/received packets time interval within a network unit

module converts the packet traces of each application into a word list by first splitting the traces by using the timing parameters and then applying the Map function to identified network units. Then, the k-TSS network language of the application is learned from its words by the Language Learner module. By applying this process for all applications, a database of k-TSS languages is obtained. To classify traffic of a system, first, the Trace Generator module extracts its sessions and their network units using the timing parameters. Then, it converts the sessions to the symbolic words to prepare the inputs of the Language Learner module. Finally, the Classifier module compares the learned language of each session of the given traffic against the languages of the database to label them. We describe each module in detail in the following.

4.1 Trace Generator

The goal of the *Trace Generator* module is to transform the packet traces of each application into traces of smaller units, while preserving the relation among these units, as the letters of a language. To do so, at first, packet traces are split into smaller units, named *network unit*, by using the three timing parameters, introduced in Sect. 3. Then, these units are categorized based on their higher-layer protocol, and each category is clustered separately and named upon it. Consequently, the names of clusters constitute the *Network Alphabet*. Finally, the network alphabets are put in together to form our traces. Therefore, the input of this subproblem is Π containing m application traces and the timing parameters st, fd, and it, while the output is a set of word list $S = \{W_1, W_2, ..., W_m\}$, where W_i is a list including the generated words of language App_i.

Network Unit Extraction. *Network Units* are extracted during three steps. At first, network traces are split into *sessions* based on the *session threshold*. Then, *flow-sessions* are extracted based on the *inactive timeout*. After that, *flow-sessions* are divided into smaller units according to the flow duration constraint. Figure 1 shows the result of applying this module on a given trace to produce sessions, flow-sessions, and network units.

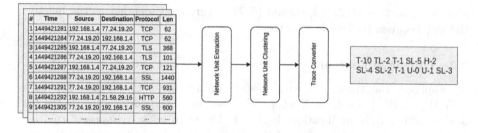

Fig. 3. Trace generation at a glance.

Network Unit Clustering. For naming network units, there is a need to group the similar ones and name them upon it. As there is no information on the number of units or their labels, an unsupervised machine learning algorithm named Kmeans++ is used for clustering these units. The only information available for each flow are those stored in the frame header of its packets. Therefore, the clustering algorithm could benefit from the knowledge of the unit higher-layer protocol. Consequently, flows having the same protocol are at worst clustered together and named upon their protocol. The packets of each flow can be further clustered in terms of multiple statistical features listed in Table 1. Two groups of features are used for clustering *network unit*, the first group consists of features number one to six, and the second group includes features number three to nine. Regarding these two groups, the statistical features group (*stats*) is another configuration of this problem.

Moreover, it is required to set the number of clusters before performing the Kmeans++ algorithm. To do so, the number of clusters is automatically determined with the help of the elbow [17] method in our approach.

Our defined *Map* function returns a concatenation of the symbol abbreviating packet protocol name and a natural number indicating its cluster number (to distinguish network units). For example, TL-5 corresponds to a packet sequence belonging to cluster 5 of the TLS protocol.

Trace Converter. In this step, first, network units are sorted based on the timestamps of their first packet. This sorting enables us to preserve the ordering relation among these units, which is ignored in previous machine learning studies. Then, the list of words $W = [w_1, w_2, ...w_n]$ is created. Each w_j equals to $\alpha_1 \alpha_2 \alpha_3 ... \alpha_{len(w_j)}$ where $len(w_j)$ shows the length of the jth word of application.

Figure 3 shows the result of applying the *Trace Generation* module where *st*, *ft*, and *fd* are set to 10, 5, and 5, respectively. For the given trace, its packets are grouped as $[(1, 2), (3, 4), (5, 7), (6), (8), (9)]$, composing network units T-10, TL-2, T-1, SL-5, H-2 and SL-4, respectively, named upon their clusters. This packet trace is split after packet 8, because of *st* constraint. Although packets 1, 2, 5, and 7 belong to a TCP flow, they are not grouped together, because of the flow duration limitation. This flow is broken between packet 2 and packet 5.

Moreover, packets $(1,2)$, $(3,4)$, and $(5,7)$ are grouped together as the inactive timeout between packets is less than ft.

4.2   Language Learner

By applying the trace generator module on Π, the set of words list $S = \{W_1, W_2, ..., W_m\}$ is achieved, where $W_i = [w_1, w_2, ...w_n]$ corresponds to the list of the words for the application $App_i \in A$. The word $w_j = \alpha_1 \, \alpha_2 \, \alpha_3 \, ... \, \alpha_{len(w_j)}$ is a finite sequence of symbols of the network alphabet, where $len(w_j)$ determine the length of the sequence w_j. To learn language $\mathcal{L}(App_i)$ as a k-testable language, we should identify its k-test vector $Z = \langle \Sigma, I, F, T \rangle$. The value of k is the second input of this subproblem.

We perform a preparation task for each word list (W_i) to handle short words. Since words with the length shorter than k $(len(w) < k)$ do not involve in producing I, T and, F sets, we attach them to the beginning of the next word in the list. Finally, $\mathcal{L}(App_i)$ is learned by scanning all words $w_j \in W_i$ with a k-window sliding parser and computing Σ, I, F, and T. As a result of executing the language learner module on word lists of all applications, the k-TSS languages database is provided.

Running Example: For the traces of Fig. 3 let $k = 3$, then, we obtain:

- $\Sigma = \{$H-2, SL-2, SL-3, SL-4, SL-5, T-1, T-10, TL-2, U-0, U-1$\}$
- $I = \{$SL-4 SL-2, T-10 TL-2$\}$
- $F = \{$SL-5 H-2, U-1 SL-3$\}$
- $T = \{$SL-2 T-1 U-0, SL-4 SL-2 T-1, T-1 SL-5 H-2, T-1 U-0 U-1, T-10 TL-2 T-1, TL-2 T-1 SL-5, U-0 U-1 SL-3$\}$

4.3   Classifier

The classifier is responsible for characterizing the given packet trace π according to k-TSS languages. Since π tends to be an interleaving of execution of multiple applications, the classifier should partition that and label each part of the π with an appropriate application.

We use the trace generator module to split it to sub-traces of π and map them as a list of words $W = [w_1, w_2, ..., w_n]$. The goal of the classifier is to label the W members.

Preparation Tasks. Some tasks are required to prepare the input trace concerning the value of k.

- Handle Short Words: this task is done similarly to what we mentioned at the Language Learner module based on the k value.
- Trim Words: Instead of sliding a k-window on a too lengthy word, we can only inspect its first symbols. Therefore, we break it down into a coefficient of k. In practice, using $4k$, we achieved acceptable accuracy (1.5% higher than when only observing $3k$ and considering $5k$ words does not increase its accuracy much more). This operation determines an upper bound time for detection.

After preparation tasks, we obtain the modified word list $W' = [w'_1, w'_2, ..., w'_n]$. Now, we learn a list of k-TSS languages for each $w' \in W'$ as $\mathcal{L}(w')$ by a k-size frame scanner. For instance, for the given $w' = $ SL-4 SL-2 T-10 TL-2 T-1 U-2, $\mathcal{L}(w')$ is obtained by k-TSS vector $Z(w')$ as $\langle \Sigma = \{$SL-2, TL-2, T-1, U-2, SL-4, T-10$\}, I = \{$SL-4 SL-2,$\}, F = \{$T-1 U-2$\}, T = \{$SL-4 SL-2 T-10, SL-2 T-10 TL-2, T-10 TL-2 T-1,TL-2 T-1 U-2$\}\rangle$.

To specify each word's class, we observe its segments instead of exploring the whole word to be more noise tolerable. To this aim, we reduce the classification problem to find the similarity between the generated k-TTS languages of the trace and the ones in the database. For each k-TSS language in the database, we calculate its proximity with $\mathcal{L}(w')$ by a *distance* function defined as:

Definition 6 (Distance Function). *Distance function D measures the proximity metric between two k-TSS languages. Let $Z = \langle \Sigma, I, F, T \rangle$ be the k-test vector of $\mathcal{L}(w')$ and $Z_i = \langle \Sigma_i, I_i, F_i, T_i \rangle$ be the k-test vector of $\mathcal{L}(App_i)$. Then, $D(\mathcal{L}(w'), \mathcal{L}(App_i))$ is computed by five auxiliary variables, measuring the sets difference fraction: $\Delta T, \Delta T_i, \Delta \Sigma, \Delta I$ and ΔF (defined in Eq. 1).*

$$\triangle T = \frac{T - T_i}{T}, \Delta T_i = \frac{T_i - T}{T_i}, \Delta \Sigma = \frac{\Sigma - \Sigma_i}{\Sigma}, \Delta I = \frac{I - I_i}{I}, \Delta F = \frac{F - F_i}{F}.$$
(1)

$$D(\mathcal{L}(w'), \mathcal{L}(App_i)) = \overline{\Delta'T} \; \overline{\Delta'T_i} \; \overline{\Delta'\Sigma} \; \overline{\Delta'I} \; \overline{\Delta'F}$$
(2)

We assume a priority among these delta metrics as $\Delta T, \Delta T_i, \Delta \Sigma, \Delta I$ and ΔF. Since T carries more information of words than I and F, we assign the highest priority to ΔT. Also, we give ΔT higher priority than ΔT_i, as T is generated from a test trace in contrast to T_i which is generated from a number of traces of an application. We specify the next priority to $\Delta \Sigma$ to take into account the alphabet sets differences. We assign the next priorities to ΔI and then ΔF, respectively, because the trimming operation of the preparation phase may impact on the end of the words while their initials are not modified. Finally, we convert the value of delta metrics from a float number in the range $[0, 1]$ to an integer number in the range $[0, 99]$, renaming them by a prime sign, for example, $\Delta'T$. By this priority, the distance function is defined as given by Eq. 2.

A word w is categorized in class j if the k-TSS language of w has the minimum distance with the k-TSS language of App_j among all the applications:

$$Class(w') = j \iff if \; D(\mathcal{L}(w'), \mathcal{L}(App_j)) = argmin_{\forall App_i \in |A|}(D(\mathcal{L}(w'), \mathcal{L}(App_i))) \quad (3)$$

Considering the running example ($\mathcal{L}(App_i)$ in Sect. 4.2 and $\mathcal{L}(w')$), the defined delta metrics are obtained as $\Delta T = 0.75$ ($\Delta'T = 74$), $\Delta T_i = 0.85$ ($\Delta'T_i = 84$), $\Delta \Sigma = 0.16$ ($\Delta'\Sigma = 16$), $\Delta I = 0$ ($\Delta'I = 0$), and $\Delta F = 1$ ($\Delta'F = 99$) and finally, the distance is computed as $D(\mathcal{L}(w'), \mathcal{L}(App_i)) = 7484160099$.

5 Evaluation

To evaluate the proposed framework, we have fully implemented it and have run multiple experiments. We evaluate our method for two classification tasks: traffic characterization and application identification. Traffic characterization deals with determining the application category of a trace such as *Chat*, and the goal of application identification is to distinguish the application of a trace such as *Skype*. To evaluate the performance of the proposed framework, we have used Precision (Pr), Recall (Rc), and F1-Measure (F1). The mathematical formula for these metrics is provided in Eq. 4.

$$Recall = \frac{TP}{TP + FN}, Precision = \frac{TP}{TP + FP}, F1 = \frac{2 * Rc * Pr}{Rc + Pr} \qquad (4)$$

5.1 Dataset

We have used the dataset of [7] to evaluate our method. This dataset is composed of captured traffic of eight applications, each executed for about 100 rounds and provided in pcap format. This set of applications comprises of Skype, VSee, TeamViewer, JoinMe, GIT, SVN, Psiphon, and Ultra, falling in four different traffic characterization categories: Chat, Remote Desktop Sharing, Version Control, and VPN. We have filtered out packets having useless information for traffic characterization and application identification tasks such as acknowledgment packets using the tshark tool[1]. To evaluate our framework, we first have randomly partitioned the dataset into three independent sets, naming train, validation, and test. These sets include 65%, 15%, and 20% of each application pcap files, respectively.

5.2 Implementation

We have fully implemented the framework with Python 3. After filtering out the unnecessary packets, 5-tuple which identifies each network flow along with timestamp and length are extracted for each packet of packet trace. Statistical features provided in Table 1 are extracted by Python Pandas libraries[2] in the *Network Unit Extraction* module (Sect. 4.1). Kmeans++ algorithm is used for the clustering purpose from the scikit-learn library[3] and StandardScaler function is used for standardizing the feature sets from the same package in the *Network Unit Clustering* module (Sect. 4.1). To automate identifying the number of clusters of each protocol, we have leveraged KneeLocator from the Kneed library[4]. For implementing the *Language Learner* module (Sect. 4.2) we benefited from k-TSS language learner part of [11] implementation and modified it based on our purpose. Finally, we have used the grid search method to choose the best value for the framework's hyper-parameters, including st, it, fd, $stats$, and k.

---

[1] https://www.wireshark.org/docs/man-pages/tshark.html.
[2] https://pandas.pydata.org/.
[3] https://scikit-learn.org/.
[4] https://pypi.org/project/kneed/.

Table 2. Framework performance

Application	Performance Metrics		
	Pr	Rc	F1
Skype	100	100	100
VSee	100	100	100
TeamViewer	87	100	93
JoinMe	95	95	95
GIT	100	100	100
SVN	100	100	100
Psiphon	95	86	90
Ultra	100	95	97
Wtd. Average	97	97	97

(a) Application Identification

Traffic Characterization	Performance Metrics		
	Pr	Rc	F1
Remote Desktop Sharing	90	97	94
VPN	97	90	94
VersionControl	100	100	100
Chat	100	100	100
Wtd. Average	97	96.5	97

(b) Traffic Characterization

5.3 Classification Results

We have used a grid search to fine-tune the parameters st, fd, it, $stats$, and k, called hyper-parameters in machine learning. Based on the average F1-Measure, consequently, the best values for the application identification task are 15 s, 5 s, 10 s, 2, and 3 for the hyper-parameters st, fd, it, $stats$, and k, respectively, while for the traffic characterization task, they are 15 s, 15 s, 10 s, 2, and 3. Using the elbow method, the number of clusters for the most used protocol such as TCP is automatically set to 23 and for a less used protocol such as HTTP is set to six, resulting in 101 total number of clusters.

We have evaluated the framework on the test set with the chosen hyper-parameters. The framework has gained weighted average F1-Measure of 97% for both tasks. Table 2 provides the proposed framework performance in terms of Precision, Recall, and F1-Measure in detail.

5.4 Comparison with Other Methods

In this section, we provide a comparison of our framework with other flow-based methods used for network traffic classification. For the application identification task, we have compared our work with [2]. In this comparison, we first have extracted 44 most used statistical flow-based features from the dataset, as the 111 initial features are not publicly available. Then, we have converted these instances to arff format to be used as Weka[5] input. To have a fair comparison, we have used the ChiSquaredAttributeEval evaluator for feature selection and finally, applied all machine learning algorithms, including J48, Random Forest, k-NN (k = 1), and Bayes Net with 10 fold cross-validation used in this work. Figure 4a compares the proposed framework performance with [2] in terms of Precision, Recall, and F1-Measure for application identification task. In most

---

[5] http://www.cs.waikato.ac.nz/ml/weka/.

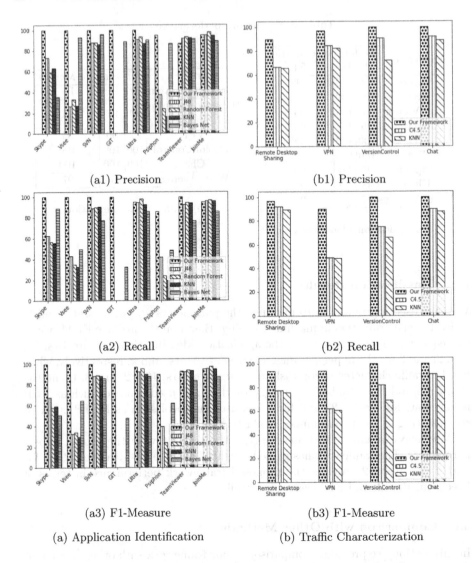

(a1) Precision   (b1) Precision

(a2) Recall   (b2) Recall

(a3) F1-Measure   (b3) F1-Measure

(a) Application Identification   (b) Traffic Characterization

Fig. 4. Performance comparison of NeTLang and statistical classifiers.

classes, our framework performed considerably better. Our framework gains much higher precision in identification of different applications in our dataset except for TeamViewer and JoinMe. In terms of recall, the work of [2] performed slightly better in detecting JoinMe and Ultra.

For the traffic characterization task, we made a comparison with the work of [1]. We have implemented a python code to extract time-related statistical flow-based features used by the authors. Using Weka, we have applied KNN and C4.5 in a 10-fold cross-validation setting which is the same as the one used in the paper. Figure 4b compares the proposed framework performance with these

Table 3. Automata learning frameworks performance and time complexity comparison. (*: Multi-thread programming will reduce the time even more)

Application	Performance Metrics						Time Complexity			
	NeTLang			Sabahi [7]			NeTLang		Sabahi [7]	
	Pr	Rc	F1	Pr	Rc	F1	Train	Test	Train	Test
Skype	100	100	100	100	81	90				
VSee	100	100	100	100	95	97				
TeamViewer	87	100	93	100	93	94				
JoinMe	95	95	95	100	81	90				
GIT	100	100	100	100	99	99	<40* min	<milisec	<20 h	<sec
SVN	100	100	100	100	99	99				
Psiphon	95	86	90	100	32	48.5				
Ultra	100	95	97	100	88	94				
Wtd. Average	97	97	97	100	83	89				

machine learning algorithms in terms of Precision, Recall, and F1-Measure for the traffic characterization task, showing that our work significantly outperforms the state-of-the-art work in application identification task.

As the proposed method in [7] is the only framework that has leveraged automata learning methods for application identification, we compare our work with it in terms of performance and time complexity (Table 3). Although the performance of [7] was computed in a two-class classification setting, NeTLang almost outperforms it in terms of Recall and F1-Measure. However, [7] has achieved higher overall precision because of its automata learning nature but owns time-consuming training and testing processes, while NeTLang considerably reduces the training time to the scale of minutes and it identifies application by observing only 4k of a network trace.

6   Related Work

Machine Learning Based Methods in Traffic Classification. Many researchers applied the machine learning techniques in traffic classification. Time-related statistical flow-based features are used in [1] to characterize traffic in the presence/absence of a virtual private network (VPN). In this work, C4.5 decision tree and K-Nearest Neighbors are applied to the extracted features to classify flows. According to their result, the C4.5 algorithm performed better with the precision of 90% for traffic characterization.

Application identification from network flows is another traffic classification task performed in [2]. They used the CfsSubsetEval and ChiSquaredAttributeEval methods in Weka to optimize the number of features and supervised learning algorithms, including J48, Random Forest, k-NN, and Bayes Net. For the dataset they used, Random Forest and k-NN had better performance in terms of accuracy (90.87% and 93.94%, respectively).

Deep Packet [3] is a recently proposed packet-level deep-learning-based framework that automatically extracts features for traffic classification. They used stacked autoencoders and one-dimensional CNN in their framework. This framework achieved a recall of 94% and 98% for traffic characterization and application identification tasks, respectively. The drawback of the deep-learning-based method is their time-consuming training process.

Automata Learning Based Methods in Traffic Classification. The most related work to ours is [7] in terms of utilizing a passive automata learning technique to derive the behavioral packet-level network models of applications. They provided a multi-steps algorithm consists of building an initial automaton and generalizing it by a state merging condition based on the behavior of well-known network protocols. Their fine granularity leads to produce large models, which make some steps of generalizing fulfill slowly for large models. Furthermore, the detection of an application was constrained to observing a complete trace of an application. Our method rectifies these shortcomings by the idea of learning a k-TSS language, which was inspired by [11]. Botnet detection has been studied in [8]. They passively learned automata for clusters of the dataset to characterize a communication profile. Other related work mainly use the automata learning for protocol modeling such as [9,10] by active automata learning, while using the desired protocol implementation as their oracle system and alphabet of the automaton derived manually by an expert.

7   Conclusion

In this paper, we have presented NeTLang, a framework which classifies network traffic by automatically extracting the network alphabet and learning the correspondent languages. To this aim, NeTLang combines an unsupervised learning algorithm and an automata learning technique to build on their strengths, to be fast and accurate and to eliminate their weaknesses, including ignoring the flow temporal relation. Moreover, we proposed a classifier by enhancing the acceptance condition of automata by machine learning with a new proximity metric. According to the experimental result, NeTLang has outperformed state-of-the-art methods used either machine learning or automata learning in traffic classification tasks by achieving the average F1-Measure of 97%. Furthermore, it has drastically decreased the learning time by approximately 96.5% in comparison with when using passive automata learning algorithm alone.

We plan to evaluate NeTLang using a public dataset. Additionally, to improve the approach performance, we will take advantage of protocols phases, naming initialization, data transmission, and finalization when extracting network units to define the alphabet more accurately.

Acknowledgments. The authors would like to thank Frits Vaandrager for his fruitful discussion on k-TSS languages and automata learning.

References

1. Draper-Gil, G., Habibi Lashkari, A., Saiful Islam Mamun, M., Ghorbani, A.A.: Characterization of encrypted and VPN traffic using time-related features. In: ICISSP (2016)
2. Yamansavascilar, B., Guvensan, M.A., Yavuz, A.G., Karsligil, M.E.: Application identification via network traffic classification. In: 2017 International Conference on Computing, Networking and Communications (ICNC), pp. 843–848, January 2017
3. Lotfollahi, M., Jafari Siavoshani, M., Shirali Hossein Zade, R., Saberian, M.: Deep packet: a novel approach for encrypted traffic classification using deep learning. Soft Comput. **24**(3), 1999–2012 (2019). https://doi.org/10.1007/s00500-019-04030-2
4. Xu, K., Zhang, Z., Bhattacharyya, S.: Profiling internet backbone traffic: behavior models and applications. SIGCOMM Comput. Commun. Rev. **35**(4), 169–180 (2005)
5. Bermolen, P., Mellia, M., Meo, M., Rossi, D., Valenti, S.: Abacus: accurate behavioral classification of P2P-TV traffic. Comput. Netw. **55**(6), 1394–1411 (2011)
6. Kinable, J., Kostakis, O.: Malware classification based on call graph clustering. J. Comput. Virol. **7**(4), 233–245 (2011)
7. Sabahi-Kaviani, Z., Ghassemi, F.: Behavioral model identification and classification of multi-component systems. Sci. Comput. Program. **177**, 41–66 (2019)
8. Hammerschmidt, C., Marchal, S., State, R., Verwer, S.: Behavioral clustering of non-stationary IP flow record data. In: 2016 12th International Conference on Network and Service Management (CNSM), pp. 297–301. IEEE (2016)
9. Fiterău-Broştean, P., Janssen, R., Vaandrager, F.: Combining model learning and model checking to analyze TCP implementations. In: Chaudhuri, S., Farzan, A. (eds.) CAV 2016. LNCS, vol. 9780, pp. 454–471. Springer, Cham (2016). https://doi.org/10.1007/978-3-319-41540-6_25
10. Fiterău-Broştean, P., Lenaerts, T., Poll, E., de Ruiter, J., Vaandrager, F., Verleg, P.: Model learning and model checking of SSH implementations. In: Proceedings of the 24th ACM SIGSOFT International SPIN Symposium on Model Checking of Software, pp. 142–151 (2017)
11. Linard, A., de la Higuera, C., Vaandrager, F.: Learning unions of k-testable languages. In: Martín-Vide, C., Okhotin, A., Shapira, D. (eds.) LATA 2019. LNCS, vol. 11417, pp. 328–339. Springer, Cham (2019). https://doi.org/10.1007/978-3-030-13435-8_24
12. McNaughton, R., Papert, S.A.: Counter-Free Automata (M.I.T. Research Monograph No. 65). The MIT Press, Cambridge (1971)
13. Garcia, P., Vidal, E., Oncina, J.: Learning locally testable languages in the strict sense. In: ALT, pp. 325–338 (1990)
14. Garcia, P., Vidal, E.: Inference of k-testable languages in the strict sense and application to syntactic pattern recognition. IEEE Trans. Pattern Anal. Mach. Intell. **12**(9), 920–925 (1990)
15. Yokomori, T., Kobayashi, S.: Learning local languages and their application to DNA sequence analysis. IEEE Trans. Pattern Anal. Mach. Intell. **20**(10), 1067–1079 (1998)
16. de la Higuera, C.: Grammatical Inference: Learning Automata and Grammars. Cambridge University Press, Cambridge (2010)
17. Thorndike, R.L.: Who belongs in the family? Psychometrika **18**, 267–276 (1953)

On the Complexity of the Upper
r-Tolerant Edge Cover Problem

Ararat Harutyunyan[1], Mehdi Khosravian Ghadikolaei[1],
Nikolaos Melissinos[1(✉)], Jérôme Monnot[1], and Aris Pagourtzis[2]

[1] Université Paris-Dauphine, PSL University, CNRS, LAMSADE,
75016 Paris, France
{ararat.harutyunyan,mehdi.khosravian-ghadikolaei,
jerome.monnot}@dauphine.fr, nikolaos.melissinos@dauphine.eu
[2] School of Electrical and Computer Engineering, National Technical University
of Athens, Polytechnioupoli, 15780 Zografou, Athens, Greece
pagour@cs.ntua.gr

Abstract. We consider the problem of computing edge covers that are
tolerant to a certain number of edge deletions. We call the problem of
finding a minimum such cover r-TOLERANT EDGE COVER (r-EC) and
the problem of finding a maximum minimal such cover UPPER r-EC. We
present several **NP**-hardness and inapproximability results for UPPER
r-EC and for some of its special cases.

Keywords: Upper edge cover · Matching · **NP**-completeness ·
Approximability

1 Introduction

In this paper we define and study *tolerant edge cover* problems. An *edge cover* of
a graph $G = (V, E)$ without isolated vertices is a subset of edges $S \subseteq E$ which
covers all vertices of G, that is, each vertex of G is an endpoint of at least one
edge in S. The *edge cover number* of a graph $G = (V, E)$, denoted by $ec(G)$, is the
minimum size of an edge cover of G and it can be computed in polynomial time
(see Chapter 19 in [29]). An edge cover $S \subseteq E$ is called *minimal* (with respect to
inclusion) if no proper subset of S is an edge cover. Minimal edge cover is also
known in the literature as an *enclaveless* set [30] or as a *nonblocker* set [14]. While
a minimum edge cover can be computed efficiently, finding the *largest minimal
edge cover* is **NP**-hard [27], where it is shown that the problem is equivalent to
finding a dominating set of minimum size. The associated optimization problem
is called *upper edge cover* (and denoted UPPER EC) [1] and the corresponding
optimal value will be denoted $uec(G)$ in this paper for the graph $G = (V, E)$.

Here, we are interested in minimal edge cover solutions tolerant to the failures
of at most $r - 1$ edges. Formally, given an integer $r \geq 1$, an edge subset $S \subseteq E$
of $G = (V, E)$ is a *tight r-tolerant edge-cover* (r-tec for short) if the deletion of

© IFIP International Federation for Information Processing 2020
Published by Springer Nature Switzerland AG 2020
L. S. Barbosa and M. Ali Abam (Eds.): TTCS 2020, LNCS 12281, pp. 32–47, 2020.
https://doi.org/10.1007/978-3-030-57852-7_3

any set of at most $r-1$ edges from S maintains an edge cover[1] and the deletion of any edge from S yields a set which is not a (tight) r-tolerant edge cover. Equivalently, we seek an edge subset S of G such that the subgraph (V, S) has minimum degree r and it is minimal with this property. For the sake of brevity we will omit the word 'tight' in the rest of the paper. Note that the case $r = 1$ corresponds to the standard notion of minimal edge cover.

As an illustrating example consider the situation in which the mayor of a big city seeks to hire a number of guards, from a security company, who will be constantly patrolling streets between important buildings. An r-tolerant edge cover reflects the desire of the mayor to guarantee that the security is not compromised even if $r-1$ guards are attacked. Providing a maximum cover would be the goal of a selfish security company, who would like to propose a patrolling schedule with as many guards as possible, but in which all the proposed guards are necessary in the sense that removing any of them would leave some building not r-covered.

Related Work. UPPER EC has been investigated intensively during recent years, mainly using the terminologies of *spanning star forests* and *dominating sets*. A *dominating set* in a graph is a subset S of vertices such that any vertex not in S has at least one neighbor in S. The *minimum dominating set problem* (denoted MINDS) seeks the smallest dominating set of G of value $\gamma(G)$. We have the equality $uec(G) = n - \gamma(G)$ [27].

Thus, using the complexity results known for MINDS, we deduce that UPPER EDGE COVER is **NP**-hard in planar graphs of maximum degree 3 [20], chordal graphs [6] (even in *undirected path graphs*, the class of vertex intersection graphs of a collection of paths in a tree), bipartite graphs, split graphs [5] and k-trees with arbitrary k [12], and it is *polynomial* in k-trees with fixed k, convex bipartite graphs [13], strongly chordal graphs [16]. Concerning the approximability, an **APX**-hardness proof with explicit inapproximability bound and a combinatorial 0.6-approximation algorithm is proposed in [28]. Better algorithms with approximation ratio 0.71 and 0.803 are given respectively in [9] and [2]. For any $\varepsilon > 0$, UPPER EDGE COVER is hard to approximate within a factor of $\frac{259}{260} + \varepsilon$ unless **P=NP** [28]. The weighted version of the problem, denoted as UPPER WEIGHTED EDGE COVER, have been recently studied in [24], in which it is proved that the problem is not $O(\frac{1}{n^{1/2-\varepsilon}})$ approximable nor $O(\frac{1}{\Delta^{1-\varepsilon}})$ in edge weighted graphs of order n and maximum degree Δ.

Related notions of dominating sets are introduced in the literature under the name r-*tuple domination* [3,18,19,21,22,25], and r-*domination* [8,18]. A set $S \subseteq V$ is called a r-*tuple dominating set* of $G = (V, E)$ if for every vertex $v \in V$, $|N_G[v] \cap S| \geq r$. The minimum cardinality of a r-tuple dominating set of G is called r-*tuple domination number* and usually denoted by $\gamma_{\times r}(G)$. The case $r = 2$ is often called *the double domination number* [21]. Complexity and

---

[1] It might be more intuitive to call such an edge cover $(r-1)$-tolerant, but for simplicity and due to the fact that each vertex has at least degree r in the cover, we have chosen to use the term r-tolerant.

approximation results on $\gamma_{\times r}(G)$ are given in [3, 25, 26] where it is proved that for any $r \geq 2$ fixed the problem is **APX**-complete in graphs of maximum degree $r + 2$ [25], **NP**-hard in split graphs and bipartite graphs for any $r \geq 1$ [26] and it admits a **PTAS** for unit disk graphs [3]. Finally, *the upper r-tuple domination number*[2] of a graph G has been recently investigated in [7] where an upper bound on this quantity for regular graphs is presented, with together a characterization of extremal graphs achieving this upper bound depending on parameter r. The particular case $r = 1$, corresponding to the *upper domination number* (denoted uds(G) here but also known as $\Gamma(G)$), has been proved **NP**-hard in [10] and extensively studied from complexity and approximability point of view in [4].

Our Contribution. In Sect. 3 we present several properties of *r-tec* solutions and of the ec$_r$ and uec$_r$ values in graphs. Furthermore, we give a characterization of *r-tec* solutions based on the $\gamma(G)$ and $\gamma_2(G)$ values of graphs, where we have equality between uec and uec$_2$ values.

In Sect. 4, we provide several complexity results. More specifically, for the DOUBLE UPPER EC we show that it is **NP**-hard in cubic planar graphs and split graphs and for the general UPPER r-EC we show that it is **NP**-hard in r-regular bipartite graphs. Furthermore, we show that the UPPER $(r + 1)$-EC is **NP**-hard in graphs with maximum degree $\Delta + 1$ if the UPPER r-EC is **NP**-hard in graphs with maximum degree Δ (and this holds even for bipartite graphs).

In Sect. 5 we present some inapproximability results starting by proving that, unless $\mathbf{P} = \mathbf{NP}$, UPPER EC is not approximable within $\frac{593}{594}$ in graphs of max degree 4 and $\frac{363}{364}$ in graphs of max degree 5 (the previous known result was for graphs with maximum degree 6). Furthermore, we present the first inaproximability results for the DOUBLE UPPER EC for graphs with maximum degree 6 and 9.

2   Definitions

Graph Notation and Terminology. Let $G = (V, E)$ be a graph and $S \subseteq V$; $N_G(S) = \{v \in V : \exists u \in S, \, vu \in E\}$ denotes the *neighborhood* of S in G and $N_G[S] = S \cup N_G(S)$ denotes the *closed neighborhood* of S. For singleton sets $S = \{s\}$, we simply write $N_G(s)$ or $N_G[s]$, even omitting G if clear from the context. The *maximum degree* and *minimum degree* of a graph are denoted $\Delta(G)$ and $\delta(G)$ respectively. For a subset of edges S, $V(S)$ denotes the vertices that are incident to edges in S. A vertex set $U \subseteq V$ induces the graph $G[U]$ with vertex set U and $e \in E$ being an edge in $G[U]$ iff both endpoints of e are in U. If $S \subseteq E$ is an edge set, then $\overline{S} = E \setminus S$, edge set S induces the graph $G[V(S)]$, while $G_S = (V, S)$ denotes the partial graph induced by S. In particular, $G_{\overline{S}} = (V, E \setminus S)$. Let also $\alpha(G)$ and $\gamma(G)$ denote the size of the largest independent and smallest dominating set of G, respectively.

An edge set S is called *edge cover* if the partial graph G_S is spanning and it is a *matching* if S is a set of pairwise non adjacent edges. An edge set S is

---
[2] The maximum cardinality of a minimal r-tuple dominating set of G.

minimal (resp., *maximal*) with respect to a graph property if S satisfies the graph property and any proper subset $S' \subset S$ of S (resp., any proper superset $S' \supset S$ of S) does not satisfy the graph property. For instance, an edge set $S \subseteq E$ is a maximal matching (resp., minimal edge cover) if S is a matching and $S + e$ is not a matching for some $e \in \overline{S}$ (resp., S is an edge cover and $S - e$ is not an edge cover for some $e \in \overline{S}$).

Problem Definitions. Let $G = (V, E)$ be a graph where the minimum degree is at least $r \geq 1$, i.e., $\delta(G) \geq r$. We assume r is a constant fixed greater than one (but all results given here hold even if r depends on the graph). A r-DEGREE EDGE-COVER[3] is defined as a subset of edges $G' = G_S = (V, S)$, such that each vertex of G is incident to at least $r \geq 1$ distinct edges $e \in S$. As r-tolerant edge-cover (or simply r-tec) we will call an edge set $S \subseteq E$ if it is a *minimal r-degree edge-cover* i.e. if for every $e \in S$, $G' - e = (V, S \setminus \{e\})$ is not an r-degree edge-cover. Alternatively, $\delta(G') = r$, and $\delta(G' - e) = r - 1$. If you seek the minimization version, all the problems are polynomial-time solvable. Actually, the case of $r = 1$ corresponds to the *edge cover* in graphs. The optimization version of a generalization of r-EC known as the MIN LOWER-UPPER-COVER PROBLEM (MIN LUCP), consists of, given a graph G where $G = (V, E)$ and two non-negative functions a, b from V such that $\forall v \in V$, $0 \leq a(v) \leq b(v) \leq d_G(v)$, of finding a subset $M \subseteq E$ such that the partial graph $G_M = (V, M)$ induced by M satisfies $a(v) \leq d_{G_M}(v) \leq b(v)$ (such a solution will be called a *lower-upper-cover*) and minimizing its total size $|M|$ among all such solutions (if any). Hence, an r-EC solution corresponds to a lower-upper-cover with $a(v) = r$ and $b(v) = d_G(v)$ for every $v \in V$. MIN LUCP is known to be solvable in polynomial time even for edge-weighted graphs (Theorem 35.2 in Chap. 35 of Volume A in [29]). We are considering two associated problems, formally described as follows.

---

r-EC
Input: A graph $G = (V, E)$ of minimum degree r.
Solution: r-tec $S \subseteq E$ of G.
Output: Minimize $|S|$.

UPPER r-EC
Input: A graph $G = (V, E)$ of minimum degree r.
Solution: r-tec $S \subseteq E$ of G.
Output: Maximize $|S|$.

For a graph G, the optimal values of r-EC and UPPER r-EC will be denoted by $\mathrm{ec}_r(G)$ and $\mathrm{uec}_r(G)$ respectively. In particular, $\mathrm{ec}_1(G) = \mathrm{ec}(G)$ and $\mathrm{uec}_1(G) = \mathrm{uec}(G)$. As indicated above, $\mathrm{ec}_r(G)$ can be computed in polynomial-time. In

---

[3] A different generalization of edge cover was considered in [17], requiring that each connected component induced by the edge cover solution contains at least r edges. Clearly, if every vertex is incident to at least r edges from the cover, then each connected component induced by the edge cover solution contains at least r edges.

Fig. 1, we illustrate the difference between the two problems r-EC and UPPER r-EC clearly for $r = 2$. Note that throughout the paper we will also use the term DOUBLE UPPER EC to refer to UPPER 2-EC.

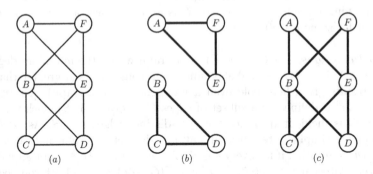

Fig. 1. Graph $G = (V, E)$ with 6 vertices and 11 edges is shown in (a); (b) and (c) give a solution for 2-EC and DOUBLE UPPER EC of size 6 and 8 respectively.

An *edge dominating set* $S \subseteq E$ of a simple graph $G = (V, E)$ is a subset S of edges such that for any edge $e \in E$ of G, at least one edge of S is incident to e. The EDGE DOMINATING SET problem (EDS in short) consists in finding an edge dominating set of minimum size; the optimal value of an edge dominating set is usually denoted eds(G). EDS is known to be **NP**-hard in general graphs (problem [GT2] in [20]). It is well known that the problem is equivalent to solve the problem consisting of finding a maximal matching of G with minimum size. According to standard terminology, this problem is also called *lower matching* (LOWER EM in short).

3   Basic Properties of r-Tolerant Solutions

The next property presents a simple characterization of feasible r-tec solution generalizing the well known result given for minimal edge covers, i.e., 1-tec, affirming that S is a 1-tec solution of G if and only if S is spanning and the subgraph (V, S) induced by S is (K_3, P_4)-free.

Property 1. Let $r \geq 1$ and let $G = (V, E)$ be a graph with minimum degree $\delta \geq r$. S is an r-tec solution of G if and only if the following conditions meet on $G_S = (V, S)$:

(1) $V = V_1(S) \cup V_2(S)$ where $V_1(S) = \{v \in V : d_{G_S}(v) = r\}$ and $V_2(S) = \{v \in V : d_{G_S}(v) > r\}$.

(2) $V_2(S)$ is an independent set of G_S.

Proof. Let $r \geq 1$ be a fixed integer and let $G = (V, E)$ be a graph instance of UPPER r-EC, i.e., a graph of minimum degree at least r. Let us prove the necessary conditions: if $S \subseteq E$ is an r-tec solution, then by construction, $V = V_1(S) \cup V_2(S)$ is a partition of vertices with minimum degree r in S. Now, if $uv \in S$ with $u, v \in V_2(S)$, then $S - uv$ is also r-tec which is a contradiction of minimality.

Now, let us prove the other direction. Consider a subgraph $G' = (V, S)$ induced by edge set S satisfying (1) and (2). By (1) it is clear G_S has minimum degree at least r. If $uv \in S$, then by (2) one vertex, say $u \in V_1(S)$ because $V_2(S)$ is an independent set. Hence, the deletion of uv leaves u of degree $r - 1$ in the subgraph induced by $G_{S \setminus \{uv\}}$ and then S is an r-tec solution.   □

Property 2. Let $r \geq 1$, for all graphs $G = (V, E)$ of minimum degree at least r, the following inequality holds:

$$2ec_r(G) \geq uec_r(G) \tag{1}$$

Proof. For a given graph $G = (V, E)$ with n vertices, let S^* be an optimal solution of UPPER r-EC, that is $|S^*| = uec_r(G)$. Let (V_1^*, V_2^*) be the associated partition related to solution S^* as indicated in Property 1. Using this characterization, we deduce $uec_r(G) \leq r|V_1^*| \leq rn$. On the other side, if G' denotes the subgraph induced by a minimum r-tec solution of value $ec_r(G)$, we get $2ec_r(G) = \sum_{v \in V} d_{G'}(v) \geq rn$. Combining these two inequalities, the results follows.   □

In particular, inequality (1) of Property 2 shows that any r-tec solution is a $\frac{1}{2}$-approximation of UPPER r-EC.

The next property is quite natural for induced subgraphs and indicates that the size of an optimal solution of a maximization problem does not decrease with the size of the graph. Nevertheless, this property is false in general when we deal with partial subgraphs; for instance, for the upper domination number, we get $uds(K_3) = 1 < 2 = uds(P_3)$. It turns out that this inequality is valid for the upper edge cover number.

Property 3. Let $G = (V, E)$ be a graph such that $0 < r \leq \delta(G)$. For every partial subgraph $G' \subseteq G$ with $\delta(G') \geq r$, the following inequality holds:

$$uec_r(G) \geq uec_r(G') \tag{2}$$

Proof. Fix an integer $r \geq 1$ and a graph $G = (V, E)$ with $\delta(G) \geq r$. Let $G' = (V', E')$ with $\delta(G') \geq r$ be a partial subgraph of G, i.e., $V' \subseteq V$ and $E' \subseteq E$. Consider an upper r-tec solution S' of G' with size $|S'| = uec_r(G')$. We prove inequality (2) by starting from $S = S'$ and by iteratively repeating the following procedure:

1. Select a vertex $v \in V$ with $d_{G_S}(v) < r$ and $e = uv \in E \setminus S$.
2. If u is covered less or more than r times by S, then $S := S + e$.
3. If vertex u is covered exactly r times by S, consider two cases:

(a) If every vertex $u' \in N_{G_S}(u)$ has degree $d_{G_S}(u') \leq r$, then $S := S + e$.
(b) Otherwise there exists a vertex $u' \in N_{G_S}(u)$ with $d_{G_S}(u') > r$ then $S := S + e - u'u$.

We repeat this process until every vertex of G is covered at least r times. Obviously, this algorithm terminates because $\sum_{v \notin V_2(S)} d_{G_S}(v)$ increase at each iteration where we recall $V_2(S) = \{v \in V : d_{G_S}(v) > r\}$. Since at each step, the size of S never decreases, we only need to show that we end the process with a r-tec solution of the whole graph. In order to prove that, we will show by induction that $V_2(S)$ is an independent set of G_S for each iteration. Using Property 1, we get an r-tec solution because $V_1(S) = V \setminus V_2(S)$. Initially $V_2(S) = V_2(S')$ is an independent set of G_S because S' is a r-tec solution of G'. Assume it is true for iteration k and consider iteration $k + 1$. During step 2, $V_2(S)$ remains unchanged in both cases $d_{G_S}(u) < r$ or $d_{G_S}(u) > r$. So by induction the result is valid. During step 3.(a), $V_2(S) := V_2(S) + u$, but $V_2(S) \cap N_{G_S}(u) = \emptyset$; so $V_2(S)$ remains an independent set of G_S. During step 3.(b), $V_2(S)$ remains unchanged if $d_{G_S}(u') > r + 1$, otherwise $V_2(S) := V_2(S) - u'$. Thus, $V_2(S)$ remains an independent set of G_S by induction. In conclusion, $\text{uec}_r(G) \geq |S| \geq |S'| = \text{uec}_r(G')$ and the property holds. □

We can prove a similar property for the size of a partial solution and the upper edge cover number. More specifically, in a graph $G = (V, E)$ a partial r-tec defined as a set $S \subseteq E$ such that the set $V_2(S) = \{v \in V \mid d_{G_S}(v) > r\}$ is an independent set of G_S.

Property 4. Let $G = (V, E)$ be a graph such that $0 < r \leq \delta(G)$. If $E' \subseteq E$ is a partial r-tec of G then the following inequality holds:

$$\text{uec}_r(G) \geq |E'| \qquad (3)$$

The proof of the above property is similar to Property's 3 except that we will start with $S := E'$.

Property 5. Let $G = (V, E)$ and $E' \subseteq E$ be a solution for UPPER r-EC of G then for any $1 \leq m < r$ we can find a set $E'' \subseteq E'$ such that E'' is a m-tec of G. Especially, if $m = r - 1$ then $E' \setminus E''$ is a tec of $G[V(E' \setminus E'')]$.

Proof. It is easy to find a m-tec of G that is a subset of E' if we start from the graph $G_{E'}$ which has minimum degree r. Any m-tec of $G_{E'}$ is a m-tec of G. Now, if $m = r - 1$ and $G' = G[V(E' \setminus E'')]$ we will prove that $E' \setminus E''$ is a tec of G'. In order to do this, we need to show that the set $V_2 = \{v \in V \mid d_{G'_{E' \setminus E''}}(v) > 1\}$ is an independent set in $G'_{E' \setminus E''}$. Let $u, v \in V_2$ be two vertices which are connected in $G'_{E' \setminus E''}$ and they have degree greater than one. Then, because E'' is a $(r-1)$-tec of G, we know that the vertices u and v are covered $r - 1$ so in E' are covered at least $r + 1$ times and they are connected. This is a contradiction because E' is an a solution for UPPER r-EC. So, if $m = r - 1$ then $E' \setminus E''$ is a tec of $G[V(E' \setminus E'')]$. □

Property 6. Let $r \geq 1$. For all graphs $G = (V, E)$ of minimum degree at least r, the following inequality holds:

$$\text{uec}_r(G) \leq r \cdot \text{uec}(G) \tag{4}$$

Proof. Let $E_1 \subseteq E$ and $E_r \subseteq E$ be the solutions of UPPER EC and UPPER r-EC. By Property 5 we can have a set $E' \subseteq E$ such that E' is a $(r-1)$-tec and $E_r \setminus E'$ is a TEC of $G[V(E_r \setminus E')]$. Then by Property 3 and because E' is a $(r-1)$-tec we have that:

$$\text{uec}_r(G) = |E'| + |E_r \setminus E'| \leq \text{uec}_{r-1}(G) + \text{uec}(G) \tag{5}$$

Therefore, by induction to r we will show that $\text{uec}_r(G) \leq r \cdot \text{uec}(G)$.
- Induction base: for $r = 2$ we have that $\text{uec}_2(G) \leq 2\text{uec}(G)$.
- Induction hypothesis: for $0 < r' < \delta(G)$ we have $\text{uec}_{r'}(G) \leq r' \cdot \text{uec}(G)$.
- Induction step: we will show that $\text{uec}_{r'+1}(G) \leq (r'+1) \cdot \text{uec}(G)$. The last can be proved using the inequality 5 and the induction hypothesis:
$\text{uec}_{r'+1}(G) \leq \text{uec}_{r'}(G) + \text{uec}(G) \leq r' \cdot \text{uec}(G) + \text{uec}(G) = (r'+1) \cdot \text{uec}(G)$. $\quad\square$

A well known relation between the domination number and the upper edge cover number for any graph G with n vertices is the following:

$$\text{uec}(G) = n - \gamma(G).$$

The previous equation can not be generalized for the r-domination number and the r upper edge cover number in any graph. However, we can prove the next relation.

Property 7. Let $r \geq 1$. For all graphs $G = (V, E)$ with n vertices and minimum degree at least r. The next inequality between $\text{uec}_r(G)$ and the r-domination number holds:

$$\text{uec}_r(G) \geq r\big(n - \gamma_r(G)\big)$$

Proof. In order to prove this we will start from a minimum r-dominating set S and we will construct a partial r-tec of G. S dominates all the vertices in $V \setminus S$ at least r times so for each $v \in V \setminus S$ we can select $u_1, \ldots, u_r \in S$ such that $vu_i \in E, \forall i = 1, \ldots, r$ and we construct the set $E_v = \{vu_i \mid i = 1, \ldots, r\}$. We will prove that the set $E' = \bigcup_{v \in V \setminus S} E_v$ is a partial r-tec. Recall that a subset E' of edges is a partial r-tec if in the graph $G_{E'}$ the set $V_2 = \{v \in V \mid d_{G_{E'}} > r\}$ is an independent set. This property holds in our set because we selected the edges in order to be incident to at least one vertex of degree exactly r. So, by the Property 4 we have that $\text{uec}_r(G) \geq |E'|$. As for the size of E' we observe that for each vertex $v \in V \setminus S$ we have used exactly r edges that are incident to a vertex of S and both of those sets ($V \setminus S$ and S) are independent in $G_{E'}$. So the size of E' is equal to $r(n - |S|)$ which implies:
$$\text{uec}_r(G) \geq |E'| = r(n - \gamma_r(G)) \qquad\qquad\square$$

In the next lemma we are interested in graphs where the equality between uec and uec_r holds.

Lemma 8. *Let $G = (V, E)$ be a graph such that $1 < r \leq \delta(G)$, then the following are equivalent:*

1. $\gamma_r(G) = \gamma(G)$
2. $\mathrm{uec}_r(G) = r \cdot \mathrm{uec}(G)$

Proof. Starting from a graph of order n were $\gamma_r(G) = \gamma(G)$ we will prove that $\mathrm{uec}_r(G) = r \cdot \mathrm{uec}(G)$. In this case we have that:

$$r \cdot \mathrm{uec}(G) = r \cdot \left(n - \gamma(G)\right) = r \cdot \left(n - \gamma_r(G)\right)$$
$$\leq \mathrm{uec}_r \qquad\qquad (\text{by Property 7})$$

In addition, we have that $r \cdot \mathrm{uec}(G) \geq \mathrm{uec}_r(G)$ (by Property 6) so we proved that:

$$\gamma_r(G) = \gamma(G) \Longrightarrow \mathrm{uec}_r = r \cdot \mathrm{uec}.$$

For the converse, let $G = (V, E)$ be a graph such that $\mathrm{uec}_r = r \cdot \mathrm{uec}$, E_r be a maximum r-tec and E_1 be a maximum tec. First suppose that there exists a partition of the vertices $V_1' \cup V_2' = V$ such that for all $v \in V_1'$ we have $d_{G_{E_r}}(v) = r$ and both of V_1', V_2' are independent in G_{E_r}. It is easy to see that V_2' is a r-dominating set of G because V_1' is independent in G_{E_r} so each vertex of it is dominated at least r times by the vertices in V_2'. Therefore, because V_1', V_2' are independent in G_{E_r} and for all $v \in V_1'$ we have $d_{G_{E_r}}(v) = r$ we can conclude that $|E_r| = r \cdot |V_1'|$. This gives us the following:

$$\mathrm{uec}_r(G) = |E_r| = r \cdot |V_1'| = r(n - |V_2'|) \leq r(n - \gamma_r(G)).$$

Because in Property 7 we showed that $\mathrm{uec}_r(G) \geq r(n - \gamma_r(G))$ we have that:

$$r(n - \gamma_r(G)) = \mathrm{uec}_r(G)r \cdot \mathrm{uec} = r(n - \gamma(G))$$

which implies that:

$$\mathrm{uec}_r = r \cdot \mathrm{uec} \Rightarrow \gamma_r(G) = \gamma(G).$$

Thus, in order to complete the proof we have to show that when the values uec and uec_r of a graph are equal then we always have a partition as the one described in the above assumption. Assuming that there is no such partition then for any partition V_1', V_2' such that V_2' is independent, all the vertices $v \in V_1'$ have degree $d_{G_{E_r}}(v) = r$ and V_1' cannot be independent. First we will show that, for the previous partition, the following property holds:

$$\text{if } v \in V_1' \text{ then } N_{G_{E_r}}(v) \nsubseteq V_1'.$$

Starting from $V_1' := V_1(E_r)$ and $V_2' := V_2(E_r)$ (see Property 1) we have to remove vertices $u \in V_1'$ such that $N_{G_{E_r}}(v) \subseteq V_1'$ until we have the desired condition. So, we repeat the following:

1. Select a vertex $u \in V_1'$ such that $N_{G_{E_r}}(u) \subseteq V_1'$.
2. Set $V_1' := V_1' - u$ and $V_2' := V_2' + u$.

We have to show that the previous process terminates when the desired properties hold for V_1' and V_2'. It is easy to see this holds because whenever we remove a $u \in V_1$ vertex such that $N_{G_{E_r}}(u) \subseteq V_1'$ from it, we reduce, at least by one, the number of vertices with this property in V_1'. Therefore, the remaining vertices $u \in V_1'$ have $d_{G_{E_r}}(u) = r$ and the vertices in V_2' are independent in G_{E_r}.

Now, due to the assumption, we know that V_1' cannot be independent (in G_{E_r}) and for all $u \in V_1'$ we have $d_{G_{E_r}}(u) = r$. This observation combined with the fact that V_2' is independent in G_{E_r} gives $r \cdot |V_1'| > E_r$. Furthermore, because each vertex in V_1' has at least one neighbor in V_2' we can select one edge for each vertex of V_1' adjacent to a vertex in V_2'. The set which consists of this edges is a partial tec of size $|V_1'|$ so by the Property 4 we have uec $\geq |V_1'|$. This gives us:

$$E_r < r \cdot |V_1'| \leq r \cdot \text{uec}$$

which is a contradiction because we have $\text{uec}_r = r \cdot \text{uec}$. $\square$

A known relation between the domination number and the r-domination number is:

$$\gamma_r(G) \geq \gamma(G) + r - 2$$

so, by the above relation and the Lemma 10 we can conclude the following:

Corollary 9. *There is no graph G such that $\text{uec}_r = r \cdot \text{uec}(G)$ for any $r \geq 3$.*

Now, we will present some graphs where the equation $\text{uec}_2 = 2\text{uec}$ holds. It is easy to see that the smallest graph in which this relation holds is C_4. Conversely, we have the following lemma.

Lemma 10. *For any graph $G = (V, E)$ with $|V| = n \geq 4$ if $\text{uec}_2(G) = 2\text{uec}(G)$ then $|E(G)| \leq \frac{n(n-2)}{2}$ when n is even and $|E(G)| \leq \frac{n(n-2)-1}{2}$ when n is odd.*

Proof. By the Lemma 8 we know that the eq. $\text{uec}_2(G) = 2\text{uec}(G)$ is equivalent to $\gamma_2(G) = \gamma(G)$. So, $\gamma(G)$ has to be greater than 2 and it deduces that the maximum number of edges in G can not be greater than $\frac{n(n-2)}{2}$, and $\frac{n(n-2)-1}{2}$ in the even and odd cases respectively. $\square$

We will now show that the above upper bounds are tight. Let $G = (V, E) = K_n$ be a clique of size n. First we construct a maximum matching $M \subseteq E$ of G; if n is even then the graph $G_{\overline{M}}$ has $\gamma_2(G_{\overline{M}}) = \gamma(G_{\overline{M}})$ and the number of edges in $G_{\overline{M}}$ is equal to $\frac{n(n-1)}{2} - |M| = \frac{n(n-1)}{2} - \frac{n}{2} = \frac{n(n-2)}{2}$. In the case when n is odd any maximum matching covers $n - 1$ vertices. Let $v \in V$ be the vertex that is not covered by the matching M. Then the only dominating set of $G_{\overline{M}}$ of size one is the $\{v\}$. Therefore if we remove any edge incident to v, let $uv \in E$ be that edge, then in $G_{\overline{M-uv}}$ we have $\gamma_2(G_{\overline{M-uv}}) = \gamma(G_{\overline{M-uv}})$ and the number of edges in $G_{\overline{M-uv}}$ are equal to $\frac{n(n-1)}{2} - |M| - 1 = \frac{n(n-1)}{2} - \frac{n+1}{2} = \frac{n(n-2)-1}{2}$.

4   Hardness of Exact Computation

In this section we provide several hardness results for the DOUBLE UPPER EC and the UPPER r-EC in some graph classes as regular graphs and split graphs.

Theorem 11. *Let $G = (V, E)$ be an $(r + 1)$-regular graph with $r \geq 2$. Then,*

$$\text{uec}_r(G) = |E| - \text{eds}(G). \tag{6}$$

Proof. In order to prove this equation, first we will show that if $S \subseteq E$ is a r-tec of G, then $\overline{S} = E \setminus S$ is an edge dominating set of G. Let (V_1, V_2) be the associated partition related to S. By the Property 1 we know that V_2 is an independent set. Because our graph is $(r + 1)$-regular, it is easy to see that $\forall v \in V_1, d_{G_S}(v) = r$ and $\forall u \in V_2, d_{G_S}(u) = r + 1$. This observation gives us that the set $\overline{S}$ covers each vertex of V_1 (they have degree r) and that all the edges in E are incident to a vertex in V_1 (because V_2 is an independent set). So, $\overline{S}$ is an edge dominating set.

Conversely, let S be a solution of EDS. We will show that there exists a r-tec of size $|\overline{S}|$. Because EDS is equivalent to LOWER EM, we consider the edge set S as a maximal matching. Now, let $V' = V \setminus V[S]$. Observe that V' is an independent set in G and each vertex $v \in V'$ has $d_{G_{\overline{S}}}(v) = r + 1$. The first holds because if there exists an edge e between two vertices of V' then $S \cup \{e\}$ will be a matching with size greater than S, which contradicts the maximality of S. The second holds because the edges in S are not incident to vertices of V' by definition, and thus, all the edges in $\overline{S}$ are incident to at least one vertex in $V[S]$. Finally, because S is a matching we have that all the vertices in $V[S]$ have degree r in $G_{\overline{S}}$ so by the Property 1, $\overline{S}$ is a r-tec. So the Eq. (6) holds.   □

Corollary 12. UPPER r-EC *is **NP**-hard in $(r + 1)$-regular bipartite graphs.*

Proof. Using the **NP**-hardness proof of EDS in r-regular bipartite graphs given in [15], the results follows from Theorem 11.   □

Corollary 13. DOUBLE UPPER EC *is **NP**-hard in cubic planar graphs.*

Proof. Using the **NP**-hardness proof of EDS for cubic planar graphs given in [23], the results follows from Theorem 11.   □

Theorem 14. UPPER $(r + 1)$-EC *is **NP**-hard in graphs of maximum degree $\Delta + 1$ if* UPPER r-EC *is **NP**-hard in graphs of maximum degree Δ, and this holds even for bipartite graphs.*

Proof. Let $G = (V, E)$ be a bipartite graph of maximum degree Δ, we construct a new graph $G' = (V', E')$ by starting from $r + 1$ copies of G. Then for each vertex $v \in V$ we add a new vertex u_v in G' and connect it to each one of the $r + 1$ copies of the vertex v.   □

Remark 15. Observe that the reverse direction of the proof applies to any $(r+1)$-tec of G'. So starting from an $(r+1)$-tec of G' we can construct an r-tec of G by checking the r-tec of the copies of G in G'. Furthermore, if we select the greater one we have the following relation:

$$|r - \text{tec}(G)| \geq \frac{1}{r+1}|(r+1) - \text{tec}(G')| - n$$

Theorem 16. DOUBLE UPPER EC *is NP-hard in split graphs.*

Proof. The proof is based on a reduction from the 2-tuple dominating set. Let $G = (V, E)$ be an instance of 2-tuple dominating set; we will construct a split graph G': First, for every vertex $v \in V$ we make three copies v^*, v', v'', and for each vertex $u \in N[v]$ we add the edges v^*u', v^*u''. Then we construct a $K_{3,2}$ and in the end we add edges in order to make a complete graph with the vertices of V^* and the only two vertices of one side of $K_{3,2}$.   □

5   Hardness of Approximation

In the following theorems we provide some inapproximability results for the UPPER EC and the DOUBLE UPPER EC.

Theorem 17. *It is NP-hard to approximate the solution of* UPPER EC *to within $\frac{593}{594}$ and $\frac{363}{364}$ in graphs of max degree 4 and 5 respectively.*

Proof. In order to prove this we will use a reduction from MIN VC. Starting from an r-regular graph $G = (V, E)$ we will construct a new graph G'; first we add a P_2 to each vertex $v \in V$. Then for each edge $e = vu \in E$ we add a new vertex v_e adjacent to v and u. In the end we remove all the starting edges E. Since MIN VC can not be approximated to within a factor $\frac{100}{99}$ (resp. $\frac{53}{52}$) in 3-regular (resp. 4-regular) graphs [11], it deduces the expected results.   □

Theorem 18. *It is NP-hard to approximate the solution of* DOUBLE UPPER EC *to within $\frac{883}{884}$ in graphs of max degree 6.*

Proof. In order to get the inapproximability result, we first make a reduction from MIN VC to UPPER EC similar to Theorem 17, then we reduce it to DOUBLE UPPER EC using a reduction proposed in Theorem 14.   □

Theorem 19. *It is NP-hard to approximate the solution of* DOUBLE UPPER EC *to within $\frac{571}{572}$ in graphs of max degree 9.*

Proof. Again, we will make a reduction from VERTEX COVER problem. Starting from a 4-regular graph $G = (V, E)$ we construct a new graph by adding a set of new vertices V_E which has one vertex v_e for each edge $e \in E$, and then adding new edges $v_e u$ if the edge e was incident to u in the original graph G. Let $G' = (V', E')$ be the new graph. It is easy to see that $|V'| = |V| + |E|$ and $\Delta(G') = 2\Delta(G) = 8$. Furthermore, we can show that from any VC of G we can construct a tec of G' of size $|TEC| = |E| + |V| - |VC|$ and conversely, from any tec of G' we can construct a VC of G of size $|VC| = |E| + |V| - |TEC|$. In order to prove the first direction we will start from a VC of G. Let S be the set of all the edges $v_e u$ where $u \in VC$. S is a partial tec of G because it covers only the vertices in $VC \cup V_E$, any vertex of V_E has degree one in G'_S and the vertices of VC are independent in G'_S. It is easy to extend S to a tec of G' by adding one edge for every vertex $v \in V \setminus VC$ that is adjacent to a vertex in VC (there exists because $v \in V$ and VC is a vertex cover of G). The extended S is a tec due to the fact that the vertices that may have greater degree than one are all in VC, which is independent in G_S and all the vertices are covered. Now we have to observe that this tec contains exactly one edge for each vertex in V_E and one for each vertex in $V \setminus VC$ so the size is exactly

$$|TEC| = |E| + |V| - |VC|.$$

Conversely, we will start from a tec of G' and we will construct a vertex cover of G of the wanted size. First we have to show that for any tec S of G', if there exists $v_e \in V_E$ such that $d_{G'_S}(v_e) = 2$ (it can not be greater because $d_{G'}(v_e) = 2$) then there exists an other tec S' of G' that has the same size and every vertex $v_e \in V_E$ has degree $d_{G_{S'}}(v_e) = 1$. This is easy to prove by repeating the following:

If there exists $e = uv \in E$ such that $d_{G_{S'}}(v_e) = 2$ then $S' := S' + v_e u - v_e v$.

This process terminates because it reduces the number of such vertices each time by one. Except that we have to show that the last set has the same size and remains a tec. Because v_e has degree two in the starting tec this means that the vertices u and v that were adjacent to it had degree one. In the new set we have degree two in vertex u and degree one in the two neighbors of it, v_e and v. So, the new set remains a tec and has the same size because we remove one and add one edge. Now, from S' we will construct a vertex cover of G. We claim that the set $U = \{v \in V \mid N_{G'_{S'}}(v) \cap V_E \neq \emptyset\}$ is a vertex cover of G of the desired size. Because for each edge $e \in E$ there exists a vertex $v_e \in V_E$, we have that U is a vertex cover of G (because it dominates the V_E). Except that, because we know that in the modified tec every vertex in V_E has degree exactly one and those edges covers only U (by construction) we need to count the edges that covers the remaining vertices. Assume that in our tec the remaining vertices $(V \setminus U)$ have degree one and are independent, then the size of our tec is

$$|TEC| = |E| + |V| - |U| \tag{7}$$

which gives us the wanted. In order to complete the reduction between VERTEX COVER and UPPER EDGE COVER we need to prove that the last assumption is always true. First observe that if two vertices $v, u \in V \setminus U$ are covered by the same edge in our tec then there exists an edge $uv = e \in E$ and a vertex $v_e \in V_E$. Then in our tec the vertex v_e must be covered by u or v which is a contradiction because non of them are in U. Now suppose that there exists vertex $v \in V \setminus U$ such that $d_{G'_{S'}}(v) \geq 2$. Because $v \notin U$ we know that there is a $u \in U$ such that

$uv \in S'$ and because u must be adjacent to a vertex in V_E in our tec this means that we have two vertices of degree at least two in a tec which is a contradiction.

After that we will do the same reduction between UPPER EC and UPPER 2-EC as in Theorem 14 so we have $uec_2(G'') = 2uec(G') + 2|V'|$ and generally by the Remark 15 we have that from any 2-tec, of value $|ec_2|$, of G'' we can construct a tec, of value $|ec|$, of G' such that the following equation holds,

$$|ec_2| \leq 2|ec| + 2|V'|. \tag{8}$$

Furthermore we have $m = |E| \leq 4|minVC|$ and $n = |V| \leq 2|minVC|$ (because we started from a 4-regular graph) which implies that from a 2-tec, of value $|ec_2|$, of G'' we can construct a vertex cover, U, of G that has the following property.

$$|ec_2| \leq 4|E| + 4|V| - 2U \tag{9}$$

The previous equation is easy to prove by the Eqs. 7, 8 and the construction of the graphs. Now we are ready to finish the proof. If we can approximate the solution of UPPER 2-EC within a factor of $1 - a$ then we can have a solution of value $|ec_2|$ such that $\frac{|ec_2|}{uec_2} \geq 1 - a$. By starting from this solution we can construct a vertex cover U for the graph G. Then we have that:

$$|U| - |minVC| \leq \frac{1}{2}(uec_2(G'') - |ec_2|) \leq \frac{1}{2}a\ uec_2(G'')$$

$$= \frac{1}{2}a\ (4m + 4n - 2|minVC|)$$

By the relations between m, n and $|minVC|$ we have the following:

$$|U| - |minVC| \leq \frac{1}{2}a\ (4m + 4n - 2|minVC|) \leq 11a\ |minVC|$$

So, we could have a $1 + 11a$ approximation for vertex cover in 4-regular graphs. Because the vertex cover is not $\frac{53}{52}$ approximable (see [11]) that gives us that a cannot be equal or less than $\frac{1}{11}\frac{1}{52}$ so the solution of DOUBLE UPPER EC is NP-hard to approximate within a factor $\frac{571}{572}$ in graphs of max degree 9. $\square$

Dedication. This paper is dedicated to the memory of our dear friend and colleague Jérôme Monnot who sadly passed away while this work was in progress.

References

1. Arumugam, S., Hedetniemi, S.T., Hedetniemi, S.M., Sathikala, L., Sudha, S.: The covering chain of a graph. Util. Math. **98**, 183–196 (2015)
2. Athanassopoulos, S., Caragiannis, I., Kaklamanis, C., Kyropoulou, M.: An improved approximation bound for spanning star forest and color saving. In: Královič, R., Niwiński, D. (eds.) MFCS 2009. LNCS, vol. 5734, pp. 90–101. Springer, Heidelberg (2009). https://doi.org/10.1007/978-3-642-03816-7_9

3. Banerjee, S., Bhore, S.: Algorithm and hardness results on Liar's dominating set and k-tuple dominating set. In: Colbourn, C.J., Grossi, R., Pisanti, N. (eds.) IWOCA 2019. LNCS, vol. 11638, pp. 48–60. Springer, Cham (2019). https://doi.org/10.1007/978-3-030-25005-8_5

4. Bazgan, C., et al.: The many facets of upper domination. Theoret. Comput. Sci. **717**, 2–25 (2018)

5. Bertossi, A.A.: Dominating sets for split and bipartite graphs. Inf. Process. Lett. **19**(1), 37–40 (1984)

6. Booth, K.S., Johnson, J.H.: Dominating sets in chordal graphs. SIAM J. Comput. **11**(1), 191–199 (1982)

7. Chang, G.J., Dorbec, P., Kim, H.K., Raspaud, A., Wang, H., Zhao, W.: Upper k-tuple domination in graphs. Discrete Math. Theoret. Comput. Sci. **14**(2), 285–292 (2012)

8. Chellali, M., Favaron, O., Hansberg, A., Volkmann, L.: k-domination and k-independence in graphs: a survey. Graphs Comb. **28**(1), 1–55 (2012)

9. Chen, N., Engelberg, R., Nguyen, C.T., Raghavendra, P., Rudra, A., Singh, G.: Improved approximation algorithms for the spanning star forest problem. Algorithmica **65**(3), 498–516 (2013)

10. Cheston, G.A., Fricke, G., Hedetniemi, S.T., Jacobs, D.P.: On the computational complexity of upper fractional domination. Discrete Appl. Math. **27**(3), 195–207 (1990)

11. Chlebík, M., Chlebíková, J.: Complexity of approximating bounded variants of optimization problems. Theoret. Comput. Sci. **354**(3), 320–338 (2006)

12. Corneil, D.G., Keil, J.M.: A dynamic programming approach to the dominating set problem on k-trees. SIAM J. Algebraic Discrete Methods 8(4), 535–543 (1987)

13. Damaschke, P., Müller, H., Kratsch, D.: Domination in convex and chordal bipartite graphs. Inf. Process. Lett. **36**(5), 231–236 (1990)

14. Dehne, F., Fellows, M., Fernau, H., Prieto, E., Rosamond, F.: NONBLOCKER: parameterized algorithmics for MINIMUM DOMINATING SET. In: Wiedermann, J., Tel, G., Pokorný, J., Bieliková, M., Štuller, J. (eds.) SOFSEM 2006. LNCS, vol. 3831, pp. 237–245. Springer, Heidelberg (2006). https://doi.org/10.1007/11611257_21

15. Demange, M., Ekim, T., Tanasescu, C.: Hardness and approximation of minimum maximal matchings. Int. J. Comput. Math. **91**(8), 1635–1654 (2014)

16. Farber, M.: Domination, independent domination and duality in strongly chordal graphs. Discrete Appl. Math. **7**, 115–130 (1984)

17. Fernau, H., Manlove, D.F.: Vertex and edge covers with clustering properties: complexity and algorithms. J. Disc. Alg. **7**, 149–167 (2009)

18. Fink, J.F., Jacobson, M.S.: n-domination in graphs. In: Graph Theory with Applications to Algorithms and Computer Science, pp. 283–300. Wiley, New York (1985)

19. Gagarin, A., Zverovich, V.E.: A generalised upper bound for the k-tuple domination number. Discrete Math. **308**(5–6), 880–885 (2008)

20. Garey, M.R., Johnson, D.S.: Computers and Intractability: A Guide to the Theory of NP-Completeness. W. H. Freeman & Co., New York (1979)

21. Harary, F., Haynes, T.W.: Double domination in graphs. Ars Comb. **55**, 201–214 (2000)

22. Haynes, T.W., Hedetniemi, S.T., Slater, P.J.: Fundamentals of Domination in Graphs, vol. 208. Pure and Applied Mathematics, Dekker, New York (1998)

23. Joseph Douglas Horton and Kyriakos Kilakos: Minimum edge dominating sets. SIAM J. Discrete Math. **6**(3), 375–387 (1993)

24. Khoshkhah, K., Ghadikolaei, M.K., Monnot, J., Sikora, F.: Weighted upper edge cover: complexity and approximability. J. Graph Algorithms Appl. **24**(2), 65–88 (2020)
25. Klasing, R., Laforest, C.: Hardness results and approximation algorithms of k-tuple domination in graphs. Inf. Process. Lett. **89**(2), 75–83 (2004)
26. Liao, C.-S., Chang, G.J.: k-tuple domination in graphs. Inf. Process. Lett. **87**(1), 45–50 (2003)
27. Manlove, D.F.: On the algorithmic complexity of twelve covering and independence parameters of graphs. Discrete Appl. Math. **91**(1–3), 155–175 (1999)
28. Nguyen, C.T., Shen, J., Hou, M., Sheng, L., Miller, W., Zhang, L.: Approximating the spanning star forest problem and its application to genomic sequence alignment. SIAM J. Comput. **38**(3), 946–962 (2008)
29. Schrijver, A.: Combinatorial Optimization: Polyhedra and Efficiency. Springer, Heidelberg (2003)
30. Slater, P.J.: Enclaveless sets and MK-systems. J. Res. Nat. Bur. Stand. **82**(3), 197–202 (1977)

Margin-Based Semi-supervised Learning Using Apollonius Circle

Mona Emadi[1(✉)] and Jafar Tanha[2(✉)]

[1] Azad University, Borujerd, Iran
emadi.mona@pnu.ac.ir
[2] Electrical and Computer Engineering Department,
University of Tabriz, Tabriz, Iran
tanha@utabrizu.ac.ir

Abstract. In this paper, we focus on the classification problem to semi-supervised learning. Semi-supervised learning is a learning task from both labeled and unlabeled data examples. We propose a novel semi-supervised learning algorithm using a self-training framework and support vector machine. Self-training is one of the wrapper-based semi-supervised algorithms in which the base classifier assigns labels to unlabeled data at each iteration and the classifier re-train on a larger training set at the next training step. However, the performance of this algorithm strongly depends on the selected newly-labeled examples. In this paper, a novel self-training algorithm is proposed, which improves the learning performance using the idea of the Apollonius circle to find neighborhood examples. The proposed algorithm exploits a geometric structure to optimize the self-training process. The experimental results demonstrate that the proposed algorithm can effectively improve the performance of the constructed classification model.

Keywords: Apollonius circle · Semi-supervised classification · Self-training · Support vector machine

1 Introduction

Typically, supervised learning methods are useful when there is enough labeled data, but in many real word tasks, unlabeled data is available. Furthermore, in practice, labeling is an expensive and time consuming task, because it needs human experience and efforts [14]. Therefore, finding an approach which can employ both labeled and unlabeled data to construct a proper model is crucial. Such a learning approach is named semi-supervised learning. In semi-supervised learning algorithms, we use labeled data as well as unlabeled data. The main goal of semi-supervised learning is to employ unlabeled instances and combine the information in the unlabeled data with the explicit classification information of labeled data to improve the classification performance. The main challenge

© IFIP International Federation for Information Processing 2020
Published by Springer Nature Switzerland AG 2020
L. S. Barbosa and M. Ali Abam (Eds.): TTCS 2020, LNCS 12281, pp. 48–60, 2020.
https://doi.org/10.1007/978-3-030-57852-7_4

in semi-supervised learning is how to extract knowledge from the unlabeled data [4,21,30]. Several different algorithms for semi-supervised learning have been introduced, such as the Expectation Maximization (EM) based algorithms [11,13,15], self-training [7,9,23], co-training [19,27], Transduction Support Vector Machine (TSVM) [1,12,26], Semi-Supervised SVM (S3VM) [3], graph-based methods [2,28], and boosting based semi-supervised learning methods [5,16,24].

Most of the semi-supervised algorithms follow two main approaches, extension of a specific base learner to learn from labeled and unlabeled examples and using a framework to learn from both labeled and unlabeled data regardless of the used base learner. Examples of the first approach include S3VM, TSVM, and LapSVM. Wrapper-based and Boosting-based methods follow the second approach, like Co-training, SemiBoost [16], MSAB [24], and MSSBoost [22]. In this article, we focus on both approaches and propose a novel semi-supervised approach based on the SVM base learner.

Support vector machine is proposed by Cortes and Vapnik [6] and is one of the promising base learners in many practical domains, such as object detection, document and web-page categorization. It is a supervised learning method based on margin as well as statistical learning theory [8]. The purpose of the support vector machine algorithm is to find an optimal hyperplane in an N-dimensional space in order to classify the data points. As shown in Fig. 1, there are many possible hyperplanes that could be selected. The optimal classification hyperplane of SVM needs not only segregating the data points correctly, but also maximizing the margin [8]. Maximizing the margin leads to a strong classification model. The standard form of SVM is only used for supervised learning tasks. This base learner can not directly handle the semi-supervised classification tasks. There are several extensions to SVM, like S3VM and TSVM. These methods use the unlabeled data to regularize the decision boundary. These methods mainly extend the SVM base classifier to semi-supervised learning, which are computationally expensive [16]. Therefore, these approaches are suitable only for small datasets.

More recently, a semi-supervised self-training has been used to handle the semi-supervised classification tasks [7,23,29]. Self-training is a wrapper-based algorithm that repeatedly uses a supervised learning method. It starts to train on labeled data only. At each step, a set of unlabeled points is labeled according to the current decision function; then the supervised method is retrained using its own predictions as additional labeled points [23]. However, the performance of this method depends on correctly predicting the labels of unlabeled data. This is important, because the selection of incorrect predictions will propagate to produce further classification errors. In general, there are two main challenges. The first is to select the appropriate candidate set of unlabeled examples to label at each iteration of the training procedure. The second is to correctly predict labels to unlabeled examples. To handle these issues, the recent studies tend to find a set of high-confidence predictions at each iteration [7,23,29]. These selected examples are typically far away from the decision boundary. Hence, this type of algorithm cannot effectively exploit information from the unlabeled data and the final decision boundary will be very close to that of the initial classifier [22].

In [29], subsets of unlabeled data are selected which are far away from the current decision boundary. Although these data have a high confident rate, they are not informative. Indeed they have little effect on the position of the hyperplane. Furthermore, adding all the unlabeled points is time consuming and may not change the decision boundary. In this paper, we propose a novel approach which tends to find a set of informative unlabeled examples at each iteration of the training procedure.

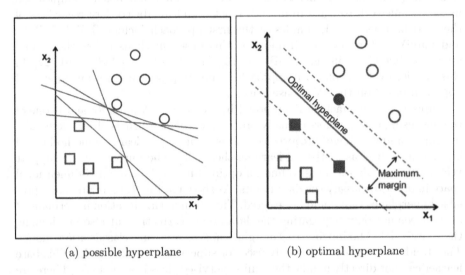

    (a) possible hyperplane   (b) optimal hyperplane

Fig. 1. Support vector machine

In our proposed approach, we select a set of unlabeled data that is potentially close to the hyperplane. In order to assign correct pseudo-label, we employ the Apollonius circle idea regarding the neighborhood examples. A Neighborhood Construction algorithm based on Apollonius Circle is used to find the neighborhood of the data points [18]. The main contributions of this paper include:

1. By selecting unlabeled data close to the decision boundary, we improve the margin.
2. We define a new method to measure the similarity between the labeled and unlabeled data points based on a geometrical structure.

We use the UCI benchmark datasets [10] to evaluate the proposed algorithm. Our experiments on a number of UCI datasets show that the proposed algorithm outperforms the state-of-the-art wrapper-based methods to semi-supervised learning.

The rest of the paper is organized as follows: Sect. 2 reviews the literature related to the proposed algorithm. Section 3, the proposed algorithm and framework are discussed. Section 4, presents the experiment results on the datasets and compare to the-state-of-the-art algorithms. Section 5, we conclude this paper.

2 Review of Concepts Related to the Density Peaks and the Apollonius Circle

In this section, the concepts of the Apollonius circle are discussed. We propose the neighborhood structure of the Apollonius circle to label the unlabeled data.

Apollonius circle is one of the most famous problems in geometry [18]. The goal is to find accurate neighborhoods. In a set of points, there is no information about the relationship between the points and some databases may not even contain topological information. It is more accurate than the Gabriel graph [25] and neighborhood graph in the neighborhood construction. It evaluates important neighborhood entirely. In this method, the first step is to find high density points and the second step is to build neighborhood groups with the Apollonius circle. The third step is analyzing points outside the radius of the Apollonian circles or points within the region.

2.1 Finding High Density Points

Rodriguez and Laio [20] presented the algorithm to find high density points (DPC). The high density points are found by this method and then stored in an array.

The points are shown by the vector $\mathbf{M} = (M_{1i}, M_{2i}, ..., M_{mi})$ where m is the number of attributes, also N_{M_i} shows k nearest neighbors of M_i. $d(M_i, M_j)$ is the Euclidean distance between M_i and M_j. The percent of the neighborhood is shown by p. The number of neighbors is obtained by $r = p \times n$, where n is the number of data points. The local density ρ_i is then defined as:

$$\rho_i = \exp\left(-\left(\frac{1}{r}\sum_{M_j \in N(M_i)} d(M_i, M_J)^2\right)\right). \tag{1}$$

$$d(M_i, M_j) = \|M_i - M_j\|. \tag{2}$$

where δ_i is the minimum distance between M_i and any other sample with higher density than p_i, which is define as below:

$$\delta_i = \begin{cases} \min_{\rho_i < \rho_j} \{d(M_i, M_j)\}, & \text{if } \exists j \quad \rho_i < \rho_j \\ \max_j \{d(M_i, M_j)\}, & otherwise \end{cases} \tag{3}$$

Peaks (high density points) are obtained using the score function. The points that have the highest score are considered as peak point.

$$score(M_i) = \delta_i \times \rho_i \tag{4}$$

In this article, number of peaks are selected based on the number of classes. Peaks are selected from the labeled set. We assign the label of peaks to the unlabeled data based on neighborhood radius of peaks. Neighboring groups are found by the Apollonius circle.

2.2   Neighborhood Groups with the Apollonius Circle

The Apollonius circle is the geometric location of the points on the Euclidean plane which have a specified ratio of distances to two fixed points A and B, this ratio is called K [17]. Apollonius circle can be seen in Fig. 2.

$$K = d_1/d_2. \tag{5}$$

The Apollonius circle based on A and B is defined as:

$$C_{AB} = \begin{cases} C_A & \text{if } K < 1 \\ C_B & \text{if } K > 1 \\ C_{inf} & \text{if } K = 1 \end{cases} \tag{6}$$

After finding the high density points, we sort these points. The peak points are indicated by $P = (P_1, P_2, ..., P_m)$ which are arranged in pairs (P_t, P_{t+1}), $t \in \{1, 2, ..., m-1\}$, the data points are denoted by $M = \{M_i | i \in \{1, 2, ..., n-m\}, M_i \notin P\}$. In the next step, data points far from the peak points are determined by the formula 7. Finally the distance of the furthest point from the peak points is calculated by the formula 8.

$$Fd_{P_t} = \max \{d(P_t, M_i) | M_i \in M \text{ and } d(P_t, M_i) < d(P_t, P_{t+1})$$

$$\text{and } d(P_t, M_i) < \min_{l=1, l \in P}^{m} d(P_l, M_i) \text{ s.t. } t \neq l \}. \tag{7}$$

$$FP_t = \{M_i | d(P_t, M_i) = Fd_t\}. \tag{8}$$

The points between the peak point and the far point are inside the Apollonius circle, circle [18].

The above concepts are used to label the unlabeled samples confidently. In our proposed algorithm, label of the peak points are assigned to the unlabeled example which are inside the Apollonius circle. The steps are shown in Fig. 3.

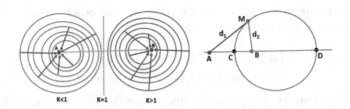

Fig. 2. Apollonius circles C_{AB} and C_B

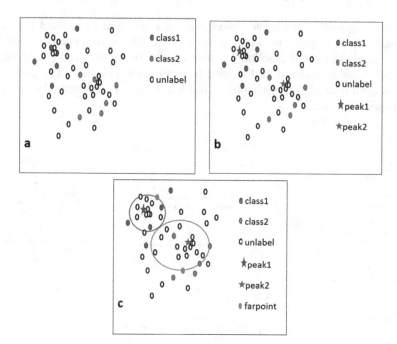

Fig. 3. Steps of making neighborhood groups with the Apollonius circle: (a) display the data distribution, (b) detecting the peak that selected from the labeled data and located in the high density region, (c) finding the far points and draw the apollonius circle

3 Our Proposed Algorithm

The proposed framework and algorithm is described in this section. The classifier works better with more labeled data, unlabeled data is much more than labeled data in semi supervised algorithms. If unlabeled data is labeled correctly and added to the labeled set, performance of the classifier improves.

Our proposed algorithm is a semi-supervised self-training algorithm. Figure 4 shows the steps of the proposed framework. In the first step, we find δ and ρ for all training data set (labeled and unlabeled) and then the high density peaks are found in the labeled data set. The number of peaks is the number of classes. Then, for each peak we find corresponding far point. Step 2 consists of two parts. One section is about selecting a set of unlabeled data which is the candidate for labeling. The distance of all unlabeled data are calculated from decision boundary. Next unlabeled data are selected for labeling that are closer to the decision boundary. Threshold (distance of decision boundary) is considered the mean distance of all unlabeled data from decision boundary.

Another part of this step is how to label unlabeled data. Our proposed method predicts the label of unlabeled data by high density peak and Apollonius circle concepts. The label of $peak_i$ is assigned to the unlabeled data that are inside the Apollonius circle corresponding $peak_i$ and in parallel the same unlabeled subset is given to SVM to label them. The agreement based on the classifier predictions and Apollonius circle is used to select a newly-labeled set The labeled set is updated and is used to retrain the classifier. The process is repeated until it reaches the stopping condition. Figure 4 represents the overview of proposed algorithm.

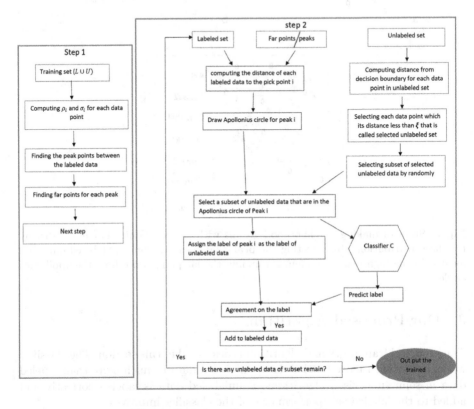

Fig. 4. Overview of proposed algorithm

4 Experiments

In this section, we perform several experiments to compare the classification performance of our proposed algorithm to the state-of-the-art semi-supervised methods using several different datasets. We also setup several experiments to show, the impact of selecting data close the decision boundary for improving the classification performance.

4.1   UCI Dataset

In the experiment, some UCI datasets are used. Table 1 summarizes the specification of 8 benchmark datasets from the UCI data repository which are used in our experiments.

4.2   Experimental Setup

For each dataset, 30% of data are kept as test set randomly and the rest are used for training set. The training set is divided into two sets of labeled data and unlabeled data. Classes are selected at a proportions equal to the original dataset for all sets. Initially, we have assigned 5% to the labeled data and 95% to the unlabeled data. We have repeated each experiment ten times with different subsets of training and testing data. We report the mean accuracy rate (MAR) of 10 times repeating the experiments.

Table 1. Summarize the properties of all the used datasets

Name	#Example	#Attributed (D)	#Class
Iris	150	4	3
Wine	178	13	3
Seeds	210	7	3
Thyroid	215	5	3
Glass	214	9	6
Banknote	1372	4	2
Liver	345	6	2
Blood	748	4	2

4.3   Results

Table 2 shows the comparison of our algorithm with some other algorithms when labeled data is 10%. The second and third columns in this table give respectively the performance of the supervised SVM and self-training SVM. The fourth column is the performance of state-of-the-art algorithm that is called STC-DPC algorithm [7]. The last column is the performance of our algorithm. Base learner for all of algorithms is SVM. Cut of distance parameter (dc) for our algorithm and STC-DPC is 0.05. From Table 2, we observe that Our algorithm works well for datasets that have a separate data density such as Iris, Seeds, Wine. Our proposed algorithm doesn't work very well if dataset is very mixed, such as banknote Fig. 5 and Fig. 6. We also investigate the behavior of the algorithms based on increasing ratio of labeled data. Fig. 7 is a comparison of the three algorithms with the increase ratio of label data from 5% to 50%.

Table 2. Experimental results of comparisons accuracy of the algorithms with 10% labeled data

Dataset	Supervised SVM	Self training	STC-DPC algorithm	Our algorithm
Iris	92.50	87	91	95.76
Wine	88.30	90.81	86.96	91.40
Seeds	84.16	74.40	81.19	92.35
Thyroid	88.95	87.21	89.65	91.72
Glass	47.44	51.15	51.15	51.93
Banknote	98.39	98.77	98.12	96.62
Liver	58.04	57.31	55.29	61.90
Blood	72.42	72.58	72.01	74.98

4.4   Impact of Selecting Data Close the Decision Boundary

In most datasets, labeling all unlabeled data can not improve the performance but also reduces the accuracy. In addition to decreasing accuracy, runtime is also increased. Although the unlabeled data that are far from the decision boundary are more reliable, they are not informative. They play little role in improving decision boundary. That's why we haven't added all the unlabeled data to the

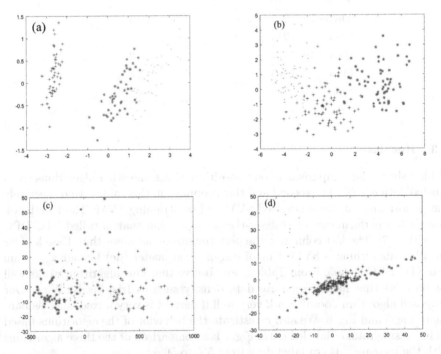

Fig. 5. Well-separated datasets: (a) iris, (b) seeds, (c) wine, (d) thyroid

training set, rather, we add those that are closer to the decision boundary than a certain value.

We show the results on a number of datasets in Table 3. The second column is the accuracy of our algorithm when we have added all the unlabeled data and the third column is the accuracy of our algorithm when we add only the data point closes to the decision boundary. As can be seen from Table 3, the accuracy of the algorithm increases when we only add unlabeled data closer to the decision boundary instead of all the points.

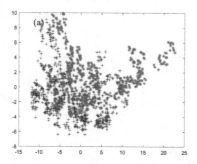

Fig. 6. Dataset with mix classes: (a) banknote

Table 3. Accuracy rate of our algorithm with all unlabeled data and near decision boundary unlabeled data

Dataset	All unlabeled data	Close unlabeled
Iris	95.59	95.76
Wine	89.18	91.26
Seeds	91.75	92.35
Thyroid	92.31	92.20
Banknote	96.58	96.62
Liver	59.85	61.90

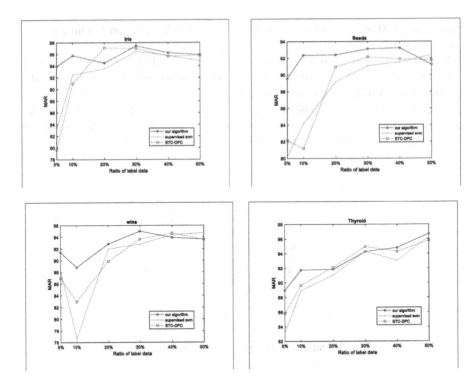

Fig. 7. Test MAR of our algorithm and supervised SVM and STC-DPC with respect to the ratio of labeled data on different datasets

5 Conclusion

In this paper, we proposed a semi-supervised self-training method based on Apollonius, named SSApolo. First candidate data are selected from among the unlabeled data to be labeled in the self training process, then, using the density peak clustering, the peak points are found and by making an Apollonius circle of each peak points, their neighbors are found and labeled. Support vector machine is used for classification. A series of experiments was performed on some datasets and the performance of the proposed algorithm was evaluated. According to the experimental results, we conclude that our algorithm performs better than STC-DPC algorithm and supervised SVM and self-training SVM, especially when classes of dataset are not very mixed. In addition, the impact of selecting data are close to decision boundary was investigated. We find that selecting data are close to decision boundary can improves the performance.

References

1. Li, Y., Wang, Y., Bi, C., Jiang, X.: Revisiting transductive support vector machines with margin distribution embedding. Knowl. Based Syst. **152**, 200–214 (2018). https://doi.org/10.1016/j.knosys.2018.04.017. http://www.sciencedirect.com/science/article/pii/S095070511830176X
2. Belkin, M., Niyogi, P., Sindhwani, V.: Manifold regularization: a geometric framework for learning from labeled and unlabeled examples. J. Mach. Learn. Res. **7**, 2399–2434 (2006)
3. Bennett, K., Demiriz, A.: Semi-supervised support vector machines. In: NIPS, pp. 368–374 (1999)
4. Chapelle, O., Schlkopf, B., Zien, A.: Semi-Supervised Learning, 1st edn. The MIT Press, Cambridge (2010)
5. Chen, K., Wang, S.: Semi-supervised learning via regularized boosting working on multiple semi-supervised assumptions. Pattern Anal. Mach. Intell. **33**(1), 129–143 (2011)
6. Cortes, C., Vapnik, V.: Support-vector networks. Mach. Learn. **20**(3), 273–297 (1995)
7. Wu, D., et al.: Self-training semi-supervised classification based on density peaks of data. Neurocomputing **275**(C), 180–191 (2018)
8. Ding, S., Zhu, Z., Zhang, X.: An overview on semi-supervised support vector machine. Neural Comput. Appl. **28**(5), 969–978 (2017). https://doi.org/10.1007/s00521-015-2113-7
9. Fazakis, N., Karlos, S., Kotsiantis, S., Sgarbas, K.: Self-trained LMT for semisupervised learning. Hindawi 2016 (2016). https://doi.org/10.1155/2016/3057481
10. Frank, A., Asuncion, A.: UCI machine learning repository (2010). http://archive.ics.uci.edu/ml
11. Habib, R., et al.: Semi-supervised generative modeling for controllable speech synthesis. ArXiv abs/1910.01709 (2019)
12. Joachims, T.: Transductive inference for text classification using support vector machines. In: ICML, pp. 200–209 (1999)
13. Langevin, M., Mehlman, E., Regier, J., Lopez, R., Jordan, M.I., Yosef, N.: A deep generative model for semi-supervised classification with noisy labels. CoRR abs/1809.05957 (2018). http://arxiv.org/abs/1809.05957
14. Li, Y., Guan, C., Li, H., Chin, Z.: A self-training semi-supervised SVM algorithm and its application in an EEG-based brain computer interface speller system. Pattern Recognit. Lett. **29**(9), 1285–1294 (2008). https://doi.org/10.1016/j.patrec.2008.01.030. http://www.sciencedirect.com/science/article/pii/S016786550800055X
15. Maaløe, L., Sønderby, C.K., Sønderby, S.K., Winther, O.: Improving semi-supervised learning with auxiliary deep generative models. In: NIPS Workshop on Advances in Approximate Bayesian Inference (2015)
16. Mallapragada, P., Jin, R., Jain, A., Liu, Y.: SemiBoost: boosting for semi-supervised learning. Pattern Anal. Mach. Intell. **31**(11), 2000–2014 (2009)
17. Partensky, M.B.: The circle of Apollonius and its applications in introductory physics. Phys. Teach. **46**(2), 104–108 (2008). https://doi.org/10.1119/1.2834533
18. Pourbahrami, S., Khanli, L.M., Azimpour, S.: A novel and efficient data point neighborhood construction algorithm based on Apollonius circle. Expert. Syst. Appl. **115**, 57–67 (2019). https://doi.org/10.1016/j.eswa.2018.07.066, http://www.sciencedirect.com/science/article/pii/S095741741830486X

19. Qiao, S., Shen, W., Zhang, Z., Wang, B., Yuille, A.L.: Deep co-training for semi-supervised image recognition. ArXiv abs/1803.05984 (2018)
20. Rodriguez, A., Laio, A.: Clustering by fast search and find of density peaks. Science **344**(6191), 1492–1496 (2014). https://doi.org/10.1126/science.1242072
21. Seeger, M.: Learning with labeled and unlabeled data (technical report). Edinburgh University (2000)
22. Tanha, J.: A multiclass boosting algorithm to labeled and unlabeled data. Int. J. Mach. Learn. Cybern. **10**(12), 3647–3665 (2019). https://doi.org/10.1007/s13042-019-00951-4
23. Tanha, J., van Someren, M., Afsarmanesh, H.: Semi-supervised self-training for decision tree classifiers. Int. J. Mach. Learn. Cybern. **8**(1), 355–370 (2015). https://doi.org/10.1007/s13042-015-0328-7
24. Tanha, J., van Someren, M., Afsarmanesh, H.: Boosting for multiclass semi-supervised learning. Pattern Recognit. Lett. **37**, 63–77 (2014)
25. Vehlow, C., Beck, F., Weiskopf, D.: The state of the art in visualizing group structures in graphs. In: EuroVis (STARs), pp. 21–40 (2015)
26. Wang, X., Wen, J., Alam, S., Jiang, Z., Wu, Y.: Semi-supervised learning combining transductive support vector machine with active learning. Neurocomput. **173**(P3), 1288–1298 (2016). https://doi.org/10.1016/j.neucom.2015.08.087. https://doi.org/10.1016/j.neucom.2015.08.087
27. Zhang, Y., Wen, J., Wang, X., Jiang, Z.: Semi-supervised learning combining co-training with active learning. Expert. Syst. Appl. **41**(5), 2372–2378 (2014). https://doi.org/10.1016/j.eswa.2013.09.035.   http://www.sciencedirect.com/science/article/pii/S0957417413007896
28. Zhou, D., Bousquet, O., Lal, T., Weston, J., Schölkopf, B.: Learning with local and global consistency. NIPS **16**, 321–328 (2004)
29. Zhou, Y., Kantarcioglu, M., Thuraisingham, B.: Self-training with selection-by-rejection. In: Proceedings of the 2012 IEEE 12th International Conference on Data Mining, ICDM 2012, pp. 795–803, December 2012. https://doi.org/10.1109/ICDM.2012.56
30. Zhu, X.: Semi-supervised learning literature survey. Technical report 1530, Computer Sciences, University of Wisconsin-Madison (2005). http://pages.cs.wisc.edu/~jerryzhu/pub/ssl_survey.pdf

Computing Boundary Cycle of a Pseudo-Triangle Polygon from Its Visibility Graph

Hossein Boomari$^{(\boxtimes)}$ and Soheila Farokhi

Department of Mathematical Sciences, Sharif University of Technology, Tehran, Iran
h.boomari1@student.sharif.ir, soheilafar.2011@gmail.com

Abstract. Visibility graph of a simple polygon is a graph with the same vertex set in which there is an edge between a pair of vertices if and only if the segment through them lies completely inside the polygon. Each pair of adjacent vertices on the boundary of the polygon are assumed to be visible. Therefore, the visibility graph of each polygon always contains its boundary edges. This implies that we have always a Hamiltonian cycle in a visibility graph which determines the order of vertices on the boundary of the corresponding polygon. In this paper, we propose a polynomial time algorithm for determining such a Hamiltonian cycle for a pseudo-triangle polygon from its visibility graph.

1 Introduction

Computing the visibility graph of a given simple polygon has many applications in computer graphics [8], computational geometry [5] and robotics [1]. There are several efficient polynomial time algorithms for this problem [5].

This concept has been studied in reverse as well: Is there any simple polygon whose visibility graph is isomorphic to a given graph and if there is such a polygon, is there any way to reconstruct it (finding positions for its vertices on the plain)? The former problem is known as recognizing visibility graphs and the latter one is known as reconstructing polygon from visibility graph. Both these problems are widely open. The only known result about the computational complexity of these problems is that they belong to *PSPACE* [3] complexity class and more precisely belong to the class of *Existence theory of reals* [7]. This means that it is not even known whether these problems are *NP-Complete* or can be solved in polynomial time. Even, if we are given the Hamiltonian cycle of the visibility graph which determines the order of vertices on the boundary of the target polygon, the exact complexity class of these polygons are still unknown.

As primitive results, these problems have been solved efficiently for special cases of tower, spiral and pseudo-triangle polygons. A tower polygon consists of two concave chains on its boundary which share one vertex and their other end points are connected by a segment (see Fig. 1a). A spiral polygon has exactly one concave and one convex chain on its boundary (see Fig. 1b). The boundary

L. S. Barbosa and M. Ali Abam (Eds.): TTCS 2020, LNCS 12281, pp. 61–71, 2020.
https://doi.org/10.1007/978-3-030-57852-7_5

of a pseudo-triangle polygon is only composed of three concave chains. The recognizing and reconstruction problems have been solved for tower polygons [2], spiral polygons [4], and pseudo-triangle polygons [6] in linear time in terms of the size of the graph. The algorithms proposed for realization and reconstruction of spiral polygon and tower polygons first find the corresponding Hamiltonian cycle of the boundary of the target polygon and then reconstruct such a polygon(if it is possible). But, the proposed algorithm for pseudo-triangle polygons needs the Hamiltonian cycle to be given as input as well as the visibility graph, and, having this pair reconstruct the target pseudo-triangle polygon. We use pseudo-triangle instead of pseudo-triangle polygon in the rest of this paper.

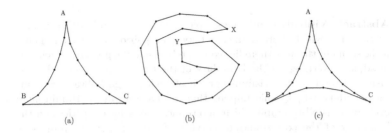

Fig. 1. a) Tower polygon, b) Spiral polygon, c) Pseudo-triangle polygon.

In this paper, we propose a method to find a Hamiltonian cycle of a realizable pseudo-triangle from its visibility graph in polynomial time. Therefore, the result of this paper in companion with the reconstruction method for realizing pseudo-triangles in [6] will solve the realization and reconstruction problems for a pseudo-triangles from its visibility graph.

In the rest of this paper, we first review the algorithm of solving recognition problem for tower polygons and give some notations, definitions and properties of pseudo-triangles to be used in next sections.

2   Preliminaries and Definitions

With a given pair of visibility graph and Hamiltonian cycle, Colley et al. proposed an efficient method to solve recognizing and reconstruction problems for tower polygons[2]. Here, we review their method, briefly.

A graph is the visibility graph of a tower polygon if and only if by removing the edges of the Hamiltonian cycle from the graph, an isolated vertex and a connected bipartite graph are obtained and the bipartite graph has *strong ordering* following the order of vertices in the Hamiltonian cycle. A strong ordering on a bipartite graph $G(V, E)$ with partitions U and W is a pair of $<_U$ and $<_W$ orderings on respectively U and W such that if $u <_U u'$, $w <_W w'$, and there are edges (u, w') and (u', w) in E, the edges (u', w') and (u, w) also exist in E. Graphs with strong ordering are also called *strong permutation graphs*.

2.1 Leveling a Tower Polygon

The algorithm proposed by Colley et al. for reconstruction of a tower polygon introduced a method named *levelling* for visibility graph of tower polygons. In this method the set of vertices of a tower polygon is covered with some subsets of its vertices called levels. The induced graph on each subset is a clique and each level is labeled with a number. It starts with level l_1 which contains the top vertex of the polygon. There are at most two candidates for the top vertex of a tower polygon. The details of the leveling method is given in Sect. 2.2. An assignment of vertices of G to the chains is called a *bordering*.

2.2 Leveling Method

Observation 1. *In $G(V,E)$ the degree of the top vertex is 2 and there are at least one and at most two vertices of degree 2. Therefore, there are at most 2 candidate for the top vertex of a tower polygon. In case there is 2 candidate for this vertex the other one is on the bottom of the tower.*

Level l_2 contains the neighbours of the top vertex. There is a set (l'_{i+1}) of at least one and at most two vertices in $V - \bigcup_{j=1}^{i} l_j$ which make a clique with vertices of l_i. l_{i+1} contains:

1. If these vertices are the last vertices of V, which are not in any levels, they make the last level (l_k).
2. Otherwise, if there is two vertices in l'_{i+1}, then $l_{i+1} = l'_{i+1}$.
3. In the last case, there is a single vertex (p) in l'_{i+1} and this vertex with one of the vertices of l_i makes l_{i+1}. In these situations exactly one of the vertices of l_i has a neighbour in $V - \bigcup_{j=1}^{i+1} l_j$. This vertex of l_i and p makes l_{i+1}.

Starting with a top vertex, there is a single leveling for the vertices of a tower polygon. The starting level contains the top vertex, the last level contains one or two vertices. If there are two vertices in the last level, they make the base of the tower and in case there is one vertex in the last level, the degree of this vertex is 2. Each of the other levels (middle levels) contains two vertices. The graph G' is the leveling graph of a visibility graph $G(V,E)$ which its vertex set is V and its edges are those edges in E which has not the endpoints in two consecutive level (see Fig. 2). We have the following observations about leveling and G'.

Observation 2. *Vertices of a middle level are not on the same chain.*

Observation 3. *G' is a bipartite graph and for each edge $(p, q) \in E(G')$, p and q do not belong to the same chain.*

Observation 4. *Any bordering that satisfies Observation 3 has a realization as tower polygon. In this realization the order of vertices of each chain follows the order of the leveling numbers.*

Consequently, the Hamiltonian cycle of the realization of each bordering is unique and is computable in $O(|E|)$.

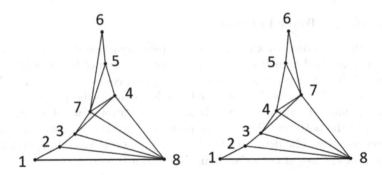

Fig. 2. Visibility graph of a tower polygon with two borderings.

Observation 5. *A tower polygon with visibility graph $G(V, E)$ and leveling graph G' with c connected components has at most 2 levelings and exactly 2^{c-1} borderings (Fig. 3).*

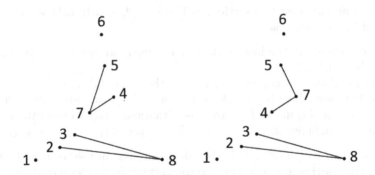

Fig. 3. Bordering graph of a tower polygon with two borderings.

2.3   Pseudo-Tower Polygons

Consider a tower polygon in which some of the bottom vertices of one of its chains are removed in such a way that the last vertices of each chain are not visible to each other. Name this kind of polygons as pseudo-tower. According to this definition, a pseudo-tower polygon has no Hamiltonian cycle. This polygon is composed of a tower polygon and an induced path at the end of a chain which can not see any vertex from the other chain (see Fig. 4). The induced path of a pseudo-tower polygons is called its *tail*.

Lemma 1. *Leveling, bordering and boundary of a visibility graph $G(V, E)$ for a pseudo-tower polygon can be computed in $O(|E(G)|)$.*

Proof. A visibility graph of a pseudo-tower polygon may have more than two vertices with two neighbours. But there is only one vertex of degree two such that its neighbours see each other. This vertex is the only candidate of the top vertex of pseudo-tower polygon. In addition, visibility graph of a pseudo-tower polygon has a single vertex p with one neighbour which is the last vertex of the tail of the polygon. Therefore, we can start at this point and find the induced path in linear time. Removing the vertices of the induced path, leaves a visibility graph which corresponds to a tower polygon which can be reconstructed in linear time.

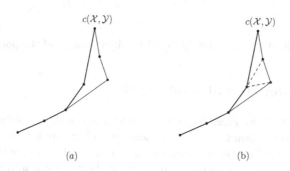

$c(\mathcal{X}, \mathcal{Y})$ $c(\mathcal{X}, \mathcal{Y})$

(a)   (b)

Fig. 4. a) Pseudo-tower polygon, b) The visibility graph of the polygon.

2.4 Pseudo-Triangle Polygons

We assume that pseudo-triangles has the layout presented in Fig. 5-a with $\mathcal{U}$, $\mathcal{V}$ and $\mathcal{W}$ as respectively its left, right and bottom concave chains, and the common vertex between two concave chains like $\mathcal{V}$ and $\mathcal{W}$ is denoted by $c(\mathcal{V}, \mathcal{W})$. The order and the name of vertices for the chains $\mathcal{U}$, $\mathcal{V}$, $\mathcal{W}$ are denoted by $<c(\mathcal{U}, \mathcal{V}), u_1, u_2, ..., c(\mathcal{U}, \mathcal{W})>$, $<c(\mathcal{U}, \mathcal{V}), v_1, v_2, ..., c(\mathcal{V}, \mathcal{W})>$ and $<c(\mathcal{U}, \mathcal{W}), w_1, w_2, ..., c(\mathcal{V}, \mathcal{W})>$, respectively. For a chain like $\mathcal{W}$ and a vertex p on this chain, the i_{th} vertex in the walk from p toward $c(\mathcal{V}, \mathcal{W})$ is denoted by $p^i_{c(\mathcal{U}, \mathcal{W})}$. The set of all vertices on chain $\mathcal{U}$ which are visible from a vertex p is denoted by $N_{\mathcal{U}}(p)$. The first and the last vertices on $\mathcal{U}$ in the walk from $c(\mathcal{U}, \mathcal{V})$ toward $c(\mathcal{U}, \mathcal{W})$, which is visible to all vertices of the set $S = p_1, ..., p_i$, are denoted by $u_{c(\mathcal{U}, \mathcal{V})}(S)$ and $U_{c(\mathcal{U}, \mathcal{V})}(S)$, respectively. In addition, we assume that that the top joint vertex $c(\mathcal{U}, \mathcal{V})$ has the least degree between the joint vertices.

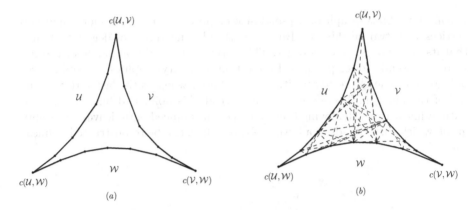

Fig. 5. a) A pseudo-triangle, b) The visibility graph of the polygon.

3   Computing Hamiltonian Cycle

For a given graph G, we present our method to find the Hamiltonian cycle $H(P)$ corresponding to the boundary cycle of some pseudo-triangle P whose visibility graph is G. We first find the vertices which can be candidates for the top vertex $c(\mathcal{U}, \mathcal{V})$. Then, we use this vertex to split the visibility graph into some regions. Finally, we introduce some necessary constraints in the visibility graph of pseudo-triangles and the detected regions, to be used to extract the Hamiltonian cycle from the visibility graph.

3.1   Find the Top Join Vertex $c(\mathcal{U}, \mathcal{V})$

As assumed before, $c(\mathcal{U}, \mathcal{V})$ has the least degree between the joint vertices.

Lemma 2. *Assume that p and q are the extreme(joint vertices) of a chain of a pseudo-triangle P and v is a non-joint vertex on this chain. The degree of v in the visibility graph of P is strictly less than the degree of exactly one of the vertices p and q.*

Proof. Without lose of generality, assume that p is a vertex on chain $\mathcal{U}$. Either all of the vertices which are visible from $c(\mathcal{U}, \mathcal{V})$ are visible from p or some vertices on chain $\mathcal{W}$ blocked their visibility. In the latter case, no vertex can block the visibility of p and the vertices which are visible from $c(\mathcal{U}, \mathcal{W})$. In addition each of the vertices which is adjacent to p on the Hamiltonian cycle is either invisible to $c(\mathcal{U}, \mathcal{V})$ or $c(\mathcal{U}, \mathcal{W})$. Therefore, the degree of p is strictly higher than one of the endpoints of its chain.

The direct result of Lemma 2 is that the degree of each non-joint vertex of a pseudo-triangle is strictly higher than one of the joint vertices of its chain. Therefore, the minimum degree of the vertices of the graph belongs to one of the joint vertices. Remind that we assume that the joint vertex with the least minimum degree in $G(P)$ is the top joint vertex of P. The degree of this vertex is strictly less than all vertices of P in $G(P)$. Therefore, to find the top vertex of P we need to find the vertex with minimum degree in $G(P)$.

Theorem 1. *In a visibility graph $G(P)$ of a pseudo-triangle P, there are at most three candidates for $c(\mathcal{U}, \mathcal{V})$ which are all joint vertices.*

Proof. As the degree of the joint vertex ($\delta(G)$) in P is strictly less than all other non-joint vertices, $G(P)$ have at most 3 vertex with degree equal to $\delta(G)$ which are the joint vertices of P.

3.2 Split the Polygon

It is simple to see that there is always a vertex w_0' on W which is visible to vertices from both chains $\mathcal{U}$ and $\mathcal{V}$. We assume that there is another vertex w_1' on W and adjacent to w_0' with this property. The degenerate cases where there is only a single vertex which is visible from both chains $\mathcal{U}$ and $\mathcal{V}$ will be handled separately in Sect. 3.4. Then, the edge $e = (w_0', w_1')$ is called a split-edge. Assume that this special edge is known. The vertices of P which are above this edge correspond to a tower polygon and we denote this polygon as $C_e(G)$. By removing the vertices of $C_e(G)$ from G, the rest of the graph will be splited into two connected components. Both of these connected components are also pseudo-tower polygons. Lets name the one which contains $c(\mathcal{U}, \mathcal{W})$ as $A_e(G)$ and the other one, which contains $c(\mathcal{V}, \mathcal{W})$, as $B_e(G)$ (see Fig. 6). Note that although we have defined these parts based on the realization of the pseudo-triangle, but, their combinatorial structures only depend on the visibility graph and the edge $e = (w_0', w_1')$.

3.3 Finding the Vertices of $C_e(G)$

There are two types of vertices in $C_e(G)$:

1. Vertices which are visible to both w_0' and w_1'. These vertices are the only vertices in $G(P)$ which have this property and we can find them in linear time. If $c(\mathcal{U}, \mathcal{V})$ belongs to this group, all vertices of $C_e(G)$ belongs to this group and the there will be no vertex in the next group.
2. Vertices which are not visible to either w_0' or w_1'. These vertices are placed above the vertices of the previous group. These vertices can be determined in linear time by using the leveling algorithm for tower polygons.

   Name the set of the vertices which are visible to both vertices of e as X, and the set of vertices which are invisible to one of the endpoints of e, Y. A vertex in pseudo-triangle polygon can see one connected part of each concave chain. Beside that, the vertices of the last level of $C_e(G)$ are visible to both vertices of e. Therefore, the leveling algorithm traverse and detect all vertices of Y before the vertices of X and we can start leveling from the top joint vertex and continue the leveling process until the leveling reach to a vertex which is visible to both

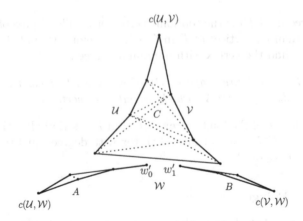

Fig. 6. Splited pseudo-triangle with respect to (w_0', w_1').

vertices of e. Name this set of vertices as C_1. Adding all vertices which are visible to both end points of e, to the set C_1 makes the set of vertices of $C_e(G)$.

After determining the vertices of $C_e(G)$ and removing them from $G(P)$, the vertices of $A_e(G)$ and $B_e(G)$ are the connected components of the remained part of GP.

3.4 Degenerate Cases

The degenerate case of pseudo-triangle polygons is the case in which there is no split-edge (there is only one vertex w_0') on the chain $\mathcal{W}$ which is visible from some vertices of both of the other chains. In this case we need another method to split the polygon. Similar to the method discussed in Sect. 3.3, we choose a split edge $e = w_0'w_1'$ in $G(P)$ and split the pseudo-triangle polygon into three parts $A_e(G)$, $B_e(G)$ and $C_e(G)$. We assume that w_1' is $w_{0c(\mathcal{V},\mathcal{W})}'^{1}$. Therefore $C_e(G)$ which contains the vertices above this split-edge contains no vertex in chain $\mathcal{V}$. The rest of process to find vertices of $C_e(G)$ and find $A_e(G)$ and $B_e(G)$ remains unchanged.

3.5 Hamiltonian Cycle of the Splited Parts

As stated before, all three parts, $A_e(G)$, $B_e(G)$, and $C_e(G)$, are tower or pseudo-tower polygons. Therefore, their recognition, reconstruction and finding their Hamiltonian cycles can be solving independently. However, Observation 5 states that there is no unique solution for the Hamiltonian cycle of a tower polygon. In addition, there are also some visibility constraints in $G(P)$ between the vertices of $A_e(G)$, $B_e(G)$, and $C_e(G)$ which have not been taken into consideration, yet. These constrains (bordering constraints) are discussed with more details as follows.

3.6   Split Constraints

The bordering graph $C'_e(G)$, which corresponds to the sub-polygon $C_e(G)$, can be computed by applying the leveling method on this visibility graph. As stated in Lemma 3 and Lemma 5, $C'_e(G)$ is a bipartite graph with some connected components. Without considering the visibility relations between vertices of $C_e(G)$ and the chain $\mathcal{W}$, each partition of these connected components can be arbitrarily assigned to each of the chains $\mathcal{U}$ and $\mathcal{V}$. Name the last vertex of $\mathcal{U}$ as p and the last vertex of $\mathcal{V}$ as q. Denote the chain of vertices of $\mathcal{W}$ which are visible from both of p and q as *common-chain*. There are two candidates for $p^1_{c(\mathcal{U},\mathcal{W})}$ in $A_e(G)$ (similarly for $q^1_{c(\mathcal{V},\mathcal{W})}$ in $B_e(G)$). Denote the chain of vertices of $\mathcal{W}$ which are visible from $p^1_{c(\mathcal{V},\mathcal{W})}$ (resp. $q^1_{c(\mathcal{V},\mathcal{W})}$) and are not on the common-chain as W' (resp. W''). In addition, denote the chain which is composed of vertices of W', W'' as *split-chain*. The following lemmas introduce two visibility constraints on bordering the vertices of $C_e(G)$ (bordering constraints):

Lemma 3. *Vertices of $\mathcal{V}$ are invisible to vertices of W'. Similarly, vertices of $\mathcal{U}$ are invisible to the vertices of W''.*

Proof. If a vertex of W' or W'' violates this condition, by the definition of the common-chain it belongs to common-chain.

Lemma 4. *For two vertices a and b on the chain $\mathcal{V}$ (resp. $\mathcal{U}$) of $C_e(G)$ such that the level of b is bigger than a, the set of vertices of $B_e(G)$ (resp. $A_e(G)$) which are visible from b are a super set of the set of vertices of $B_e(G)$ (resp. $A_e(G)$) which are visible from a.*

Proof. The vertices of the chain $\mathcal{V}$ and the set of vertices of $\mathcal{W}$ which are visible from some vertex in $\mathcal{V}$ make a pseudo-tower polygon. The statement of the lemma holds in every pseudo-tower polygon.

For a vertex a on chain $\mathcal{U}$ in $C_e(G)$ the vertex with the highest level in $\mathcal{V}$ is the vertex which blocked the sight of a and the invisible vertices (blocking vertex of a) of W''. Therefore, we have:

Lemma 5. *For every vertex a on chain $\mathcal{U}$ (resp. $\mathcal{V}$) in $C_e(G)$, the set of vertices of $B_e(G)$ (resp. $A_e(G)$) which are visible to the blocking vertex of a is a super set of the set of vertices of $B_e(G)$ (resp. $A_e(G)$) which are visible to a.*

Verifying each of these constraints for a boardering needs $O(E)$ time. In addition these constraints force that two different borderings which satisfy these constraints to be isomorph to each other, because if a pair of vertices are arbitrarily placed on $\mathcal{V}$ or $\mathcal{W}$ then they must have the same neighbours in $A_e(G)$ and $B_e(G)$ and also the same neighbours in split-chain (otherwise, swapping the chain of the vertices violates one of the increasing pattern in either Lemma 4 or Lemma 5). Therefore, there is only constant number of non-isomorph borderings which satisfy these three linear time verifiable constraints. To find this bordering we can begin with an arbitrary bordering of $C_e(G)$ and sweep the vertices of one of its chains from top to bottom. If a constraint is violated on a vertex like a, the chain of vertices of the bordering graph, G', which are in the same component with a must be swapped so that the bordering graph remains bipartite and have a chance to satisfy the bordering constraints. By continuing these constraints, it will either iterate over all vertices of $C_e(G)$ or it violates one of the constraints. Therefore we have the following theorem:

Theorem 2. *There are linear time verifiable constraints (bordering constraints) such that $O(1)$ number of non-isomorph pairs (G, H) for a pseudo-triangle with visibility graph G, Hamiltonian cycle H, the top vertex $c(\mathcal{U}, \mathcal{V})$ and the split-edge e, satisfies them.*

3.7   Recognizing the Visibility Graph

Theorem 2 leads to the final step of our method. We have a visibility graph G with a Hamiltonian cycle H and we need to verify that whether there is a pseudo-triangle with this pair as its visibility graph and Hamiltonian cycle. As stated in Sect. 1, this problem can be solved in $O(E)$. In the next section we review the outline and computational complexity of our method and prove that all the steps are polynomial time computable.

4   Computational Complexity

Our method to find Hamiltonian cycle of a visibility graph G of a pseudo-triangle has these steps:

1. Determine the candidates of the top joint vertex of the polygon. Lemma 2 states that this step takes $O(E)$ time and obtains at most three candidates for the top joint vertex.
2. Determine the split-edge of the polygon. This step chooses at most $O(E)$ candidates for split-edge.
3. Split the polygon using the determined split-edge. In this step, we find the vertices of the sub-polygon C and split the rest of the graph into its two connected components. The induced graph on the vertices of each connected component produces the sub-polygons B and C. This can be computed in $O(E)$ time.
4. Compute the leveling and G' of each sub-polygon and determine the Hamiltonian cycle of each sub-polygon. The leveling of each sub-polygon is computable in $O(E)$ time using the leveling algorithm for tower polygons.
5. Determine the split-chain. There are $O(1)$ choices for the last vertex of each of the chains of each sub-polygon. This takes $O(E)$ time to determine these candidates from the leveling graph. After determining these vertices, it takes $O(E)$ time to find the complete split-chain.
6. Compute the Hamiltonian cycle of the graph. It takes $O(E)$ time to apply the constraints on the leveling graph and find the Hamiltonian cycle (bordering) of the pseudo-triangle.

7. Verify the pair of visibility graph and Hamiltonian cycle. It takes $O(E)$ time to solve the recognition problem of pseudo-triangle for a pair of visibility graph and Hamiltonian cycle.

In summary, this method takes a graph G and spends $O(E^2)$ time to determine $O(E)$ pairs of (G, H) as the total possible candidates for the visibility graph and Hamiltonian cycle of a pseudo-triangle whose visibility graph and G are isomorph. It takes $O(E)$ time to solve the recognition problem for each of these pairs. Then, we have the final theorem:

Theorem 3. *Having only the visibility graph $G(V, E)$, recognizing and reconstruction problems for pseudo-triangles can be solved in $O(E^2)$.*

References

1. Belta, C., Isler, V., Pappas, G.J.: Discrete abstractions for robot motion planning and control in polygonal environments. IEEE Trans. Robot. **21**(5), 864–874 (2005)
2. Colley, P., Lubiw, A., Spinrad, J.: Visibility graphs of towers. Comput. Geom. **7**(3), 161–172 (1997)
3. Everett, H.: Visibility graph recognition - P.hD. thesis (1990)
4. Everett, H., Corneil, D.G.: Recognizing visibility graphs of spiral polygons. J. Algorithms **11**(1), 1–26 (1990)
5. Ghosh, S.K.: Visibility Algorithms in the Plane. Cambridge University Press, Cambridge (2007)
6. Mehrpour, S., Zarei, A.: Pseudo-triangle visibility graph: characterization and reconstruction (2019)
7. Richter-Gebert, J.: Mnëv's universality theorem revisited. Séminaire Lotaringien de Combinatorie (1995)
8. Teller, S., Hanrahan, P.: Global visibility algorithms for illumination computations. In: Proceedings of the 20th Annual Conference on Computer Graphics and Interactive Techniques, pp. 239–246. ACM (1993)

Improved Algorithms for Distributed Balanced Clustering

Kian Mirjalali and Hamid Zarrabi-Zadeh[✉]

Sharif University of Technology, Tehran, Iran
mirjalali@ce.sharif.edu, zarrabi@sharif.edu

Abstract. We study a weighted balanced version of the k-center problem, where each center has a fixed capacity, and each element has an arbitrary demand. The objective is to assign demands of the elements to the centers, so as the total demand assigned to each center does not exceed its capacity, while the maximum distance between centers and their assigned elements is minimized. We present a deterministic $O(1)$-approximation algorithm for this generalized version of the k-center problem in the distributed setting, where data is partitioned among a number of machines. Our algorithm substantially improves the approximation factor of the current best randomized algorithm available for the problem. We also show that the approximation factor of our algorithm can be improved to $5 + \varepsilon$, when the underlying metric space has a bounded doubling dimension.

1 Introduction

Clustering is a well-known problem with various applications. The problem is in particular important in distributed environments, where we are dealing with big amounts of data. In these settings, no single machine can store the whole data, and hence, data in partitioned among several nodes.

The k-center problem is a popular formulation of clustering, consisting of a set S of n elements in a metric space (U, d), and an integer k. The objective is to select k elements from S as centers and assign each element of S to one of the centers, so as the maximum distance between elements and their assigned centers is minimized. Naturally, each element is assigned to its nearest center.

In the *balanced k-center* problem (also known as *capacitated k-center*), each center has a fixed capacity L, bounding the number of elements that can be covered by that center. Obviously, when centers have capacity, elements are not necessarily covered by their nearest centers. The *weighted balanced k-center* is a generalization of this problem where each element x has a weight/demand $w(x)$. When an element x is assigned to one center, it uses $w(x)$ units of the center's capacity. However, an element can be assigned to more than one center in general, each covering a portion of its demand. The problem in its general form has natural applications in facility location scenarios where resources have capacities, and users have specific demands.

© IFIP International Federation for Information Processing 2020
Published by Springer Nature Switzerland AG 2020
L. S. Barbosa and M. Ali Abam (Eds.): TTCS 2020, LNCS 12281, pp. 72–84, 2020.
https://doi.org/10.1007/978-3-030-57852-7_6

Related Work. The k-center problem is known to be NP-hard. The *furthest-point* greedy algorithm proposed by Gonzalez [14], yields a 2-approximate solution. Hochbaum and Shmoys [15] proposed another 2-approximation algorithm based on parametric pruning. It is known that no $2-\varepsilon$ approximation algorithms is possible for the k-center problem, unless P = NP.

The capacitated version of k-center was first introduced by Barilan *et al.* [3]. They presented a 10-approximation for the problem. Khuller and Sussmann 6 improved the approximation factor to 6, and showed that the factor can be further improved to 5 for soft capacities, i.e., when elements can be selected as center more than once. Cygan *et al.* [8] studied a non-uniform variant of the problem, where each element has a specific capacity if selected as center, and presented an $O(1)$-approximation algorithm for this problem. The constant factor was later improved to 9 by An *et al.*[2]. Other variants of the capacitated k-center problem have been also studied in the literature (see, e.g., [6,9–11,13]).

The k-center problem is also studied in the distributed environments when dealing with big data. Several variations and approximation algorithms have been proposed in this context [5,12,16,19–21]. Bateni *et al.* [4] studied the balanced version of the problem in the distributed environment. and presented a randomized algorithm with approximation guarantee 32β, where β is the approximation factor of the corresponding centralized algorithm for weighted balanced k-center. Using the current best bound of $\beta = 5$ [18], their algorithm yields an approximation factor of 160.

Our Results. We present a new deterministic approximation algorithm for the weighted balanced k-center problem in distributed environments, achieving an approximation guarantee of $9\beta + 4$, where β is the approximation factor of the corresponding centralized algorithm for weighted balanced k-center. This substantially improves the current best approximation factor of 32β due to Bateni *et al.* [4]. Our algorithm can be implemented in constant number of rounds in massively parallel computation (MPC) models, such as MapReduce. Moreover, our algorithm uses a small amount of communication, which we show is optimal under fair assumptions. We further show that the approximation factor of our algorithm can be improved to $5+\varepsilon$ if the underlying metric space has a bounded doubling dimension.

Our algorithm uses the "composable coresets" framework introduced by Indyk *et al.* [17]. In this framework, a small subset of data (so-called a coreset) is carefully extracted from each machine, in such a way that the union of coresets contains a good approximation of the whole data set. This framework has been successfully used to devise approximation algorithms for several other optimization problems in distributed settings (see, e.g., [1,7,22,23]).

The rest of this paper is organized as follows. In Sect. 2, the basic definitions and formulation of the problem is given. In Sect. 3, we describe our distributed approximation algorithm for the weighted balanced k-center problem and analyze its approximation factor. In Sect. 4, we show how the approximation factor of our algorithm can be improved to $5+\varepsilon$ in metric spaces with bounded doubling dimension.

2   Preliminaries

Given two sets A and B, a *relation* from A to B is a subset of the Cartesian product of A to B. We can generalize this concept by adding a multiplicity to each pair $a \in A$ and $b \in B$, denoting the weight of relation between a and b.

Definition 1. *Given two sets A and B, a weighted relation R from A and B is a function $R : A \times B \longrightarrow \mathbb{N}_0$, where $R(a, b)$ stands for the number of relationships between $a \in A$ and $b \in B$. In particular, $R(a, b) = 0$ means that there is no relation between a and b.*

In the weighted balanced k-center problem, we are given as input a set S of n elements in a metric space (U, d), an integer k denoting the number of centers, and an integer L representing the capacity of each center. For each element $x \in S$, a demand (weight) $w(x)$ is also given, representing the number of times required for x to be covered by centers. A *clustering* $C = (D, A)$ consists of a set $D = \{c_1, c_2, \ldots, c_k\}$ of (not-necessarily distinct) centers selected from S, and a weighted relation A from S to D representing the assignment of elements to the centers. More precisely, $A(x, c_i)$ represents the number of times $x \in S$ is covered by center c_i. A clustering $C = (D, A)$ is *feasible*, if the following two conditions hold:

$$\forall x \in S : \sum_{1 \leq i \leq k} A(x, c_i) = w(x),$$

and

$$\forall c_i \in D : \sum_{x \in S} A(x, c_i) \leq L.$$

The latter is called *capacity constraint*, and the former is called *demand constraint*. Note that we are considering the soft version of the problem, where an element can be selected as a center more than once. The objective of the weighted balanced k-center problem is to find a feasible clustering $C = (D, A)$ minimizing the cost

$$R_{S,w}(C) := \max_{\substack{x \in S, \, c_i \in D \\ A(x, c_i) > 0}} d(x, c_i).$$

Obviously, the problem has no feasible solution if $\sum_{x \in S} w(x) > kL$.

3   Distributed Weighted Balanced k-Center

In this section, we present our distributed algorithm for the weighted balanced k-center problem. We assume that the input data set S is partitioned into m subsets $S_1, S_2, \ldots, S_m$, each stored in a separate machine. A good point about our algorithm is that we have no specific assumption on the partitioning of data, such as a particular ordering or a random partitioning.

Our algorithm uses the following two centralized approximation algorithms as subroutines:

– Algorithm $\mathcal{A}$: an α-approximation algorithm for the k-center problem,
– Algorithm $\mathcal{B}$: a β-approximation algorithm for the weighted balanced k-center problem.

The pseudo-code of our algorithm is presented in Algorithm 1. In this algorithm, we first run algorithm $\mathcal{A}$ separately in each machine to obtain m coresets each of size k. The coresets are then composed into a single set T in the central machine, and algorithm $\mathcal{B}$ is applied on this set to obtain a set D of k centers, along with its corresponding assignment. The set D is then sent back to the machines to obtain the final assignment. A general schema of the algorithm is illustrated in Fig. 1.

Algorithm 1. DISTRIBUTED WEIGHTED BALANCED k-CENTER

Input: Data sets $S_1, \ldots, S_m$, an integer k, a capacity L, and a weight function w
Output: A feasible k-clustering of the set $S = \bigcup_{i=1}^{m} S_i$
1: For each $1 \leq i \leq m$, run algorithm $\mathcal{A}$ on S_i to obtain a clustering $C_i = (D_i, A_i)$.
2: For each center $c \in D_i$, define $w'(c) = \sum_{x \in S_i} A_i(x, c)$
3: Send D_i's and their demands w' to the central machine. Let $T = \bigcup_{i=1}^{m} D_i$.
4: Run algorithm $\mathcal{B}$ on $\langle T, w' \rangle$ to obtain a clustering $C = (D, A)$.
5: Send C back to the machines containing S_i's.
6: In machine i, assign demands covered by $c \in D_i$ to the centers in D based on A. Call the new assignment A'.
7: **return** clustering $C' = (D, A')$.

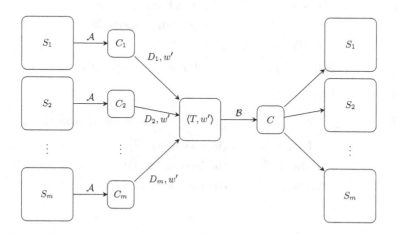

Fig. 1. A general schema of Algorithm 1.

Clearly, if the runtime of the algorithms $\mathcal{A}$ and $\mathcal{B}$ is polynomial, the whole algorithm runs in polynomial time. Now we analyze the approximation factor of Algorithm 1. In the following, we assume that $C^* = (D^*, A^*)$ is an optimal solution for the whole input set $S = \bigcup_{i=1}^{m} S_i$.

Lemma 1. *The cost of clustering computed for each subset in the first step of Algorithm 1 is not greater than 2α times the cost of optimal solution for the whole set. In other words, for all $1 \leq i \leq m$:*

$$R_{S_i,w}(C_i) \leq 2\alpha R_{S,w}(C^*).$$

Proof. Let $C^\dagger$ and $C_i^\dagger$ be optimal solutions for the uncapacitated k-center problem on inputs S and S_i, respectively. As the uncapacitated k-center problem is a relaxed form of its capacitated version, we have:

$$R_{S,w}(C^\dagger) \leq R_{S,w}(C^*) \tag{1}$$

Now, we build a feasible solution $\hat{C} = (\hat{D}, \hat{A})$ for the uncapacitated k-center problem on input S_i, based on $C^\dagger = (D^\dagger, A^\dagger)$ in the following way. If $D^\dagger = \{q_1, \ldots, q_k\}$, we set $\hat{D} = \{\hat{q}_1, \ldots, \hat{q}_k\}$, where $\hat{q}_j$ be the nearest member of S_i to q_j. Therefore:

$$\forall t \in S_i : d(q_j, \hat{q}_j) \leq d(q_j, t) \tag{2}$$

We also define $\hat{A}(x, \hat{q}_j)$ the same as $A^\dagger(x, q_j)$. Since $C_i^\dagger$ is the optimal solution and $\hat{C}$ is a feasible solution for the uncapacitated k-center problem on input S_i:

$$R_{S_i,w}(C_i^\dagger) \leq R_{S_i,w}(\hat{C}) \tag{3}$$

Assume that element $x \in S_i$ has the maximum distance from its covering center $\hat{q}_j$ in $\hat{C}$. Thus:

$$
\begin{aligned}
R_{S_i,w}(\hat{C}) &= d(x, \hat{q}_j) && \text{(by definition of } R_{S_i,w}) \\
&\leq d(x, q_j) + d(q_j, \hat{q}_j) && \text{(by triangle inequality)} \\
&\leq d(x, q_j) + d(q_j, x) && \text{(by (2))} \\
&\leq 2R_{S,w}(C^\dagger) && (\hat{A}(x, \hat{q}_j) > 0 \implies A^\dagger(x, q_j) > 0) \quad (4)
\end{aligned}
$$

Putting all together, we get:

$$
\begin{aligned}
R_{S_i,w}(C_i) &\leq \alpha R_{S_i,w}(C_i^\dagger) && \text{(approximation factor of algorithm } \mathcal{A}) \\
&\leq \alpha R_{S_i,w}(\hat{C}) && \text{(by inequality (3))} \\
&\leq 2\alpha R_{S,w}(C^\dagger) && \text{(by inequality (4))} \\
&\leq 2\alpha R_{S,w}(C^*) && \text{(by inequality (1))}
\end{aligned}
$$

$\square$

Lemma 2. *In Algorithm 1, the cost of clustering C computed by algorithm $\mathcal{B}$ for input $\langle T, w' \rangle$ is not greater than $\beta(4\alpha + 1)$ times the cost of optimal solution for the whole input. In other words:*

$$R_{T,w'}(C) \leq \beta(4\alpha + 1) R_{S,w}(C^*).$$

Proof. Algorithm $\mathcal{A}$ used in the first phase of Algorithm 1 solves an uncapacitated version of k-center, and thus, each element of S is assigned by $\mathcal{A}$ to a single center in T. We can model this assignment with a function $f : S \longrightarrow T$ and its reverse $F : T \longrightarrow P(S)$, so that for each $x \in S$, $f(x)$ is the covering center of x, and for each $t \in T$, $F(t)$ is the set of elements covered by t. Therefore, the demand $w'(t)$ for $t \in T$ (computed in step 2 of Algorithm 1) can be written as

$$w'(t) = \sum_{x \in F(t)} w(x). \tag{5}$$

Let $D^* = \{q_1^*, q_2^*, \ldots, q_k^*\}$ be the set of centers in an optimal clustering $C^* = (D^*, A^*)$ for the weighted balanced k-center problem on the whole input $\langle S, w \rangle$. We build a feasible solution $\hat{C} = (\hat{D}, \hat{A})$ for the same problem on input $\langle T, w' \rangle$ using C^* and functions f and F. The set of centers in $\hat{C}$ is $\hat{D} = \{\hat{q}_1, \ldots, \hat{q}_k\}$ where $\hat{q}_j$ is $f(q_j^*)$ (for $1 \leq j \leq k$), and the coverage assignment $\hat{A} : T \times \hat{D} \longrightarrow \mathbb{N}_0$ is defined as

$$\hat{A}(t, \hat{q}_j) := \sum_{x \in F(t)} A^*(x, q_j^*) \qquad (\text{for } t \in T, 1 \leq j \leq k).$$

To prove the feasibility of $\hat{C}$, we only need to verify that its coverage assignment satisfies the following two constraints:

- Demand constraint: for each element $t \in T$, we have:

$$\sum_{1 \leq j \leq k} \hat{A}(t, \hat{q}_j) = \sum_{1 \leq j \leq k} \sum_{x \in F(t)} A^*(x, q_j^*) \quad (\text{by definition of } \hat{A})$$

$$= \sum_{x \in F(t)} \sum_{1 \leq j \leq k} A^*(x, q_j^*)$$

$$= \sum_{x \in F(t)} w(x) \qquad (\text{due to demand constraint in } C^*)$$

$$= w'(t) \qquad (\text{by (5)})$$

- Capacity constraint: for each center $\hat{q}_j$ $(1 \leq j \leq k)$, we have:

$$\sum_{t \in T} \hat{A}(t, \hat{q}_j) = \sum_{t \in T} \sum_{x \in F(t)} A^*(x, q_j^*) \quad (\text{by definition of } \hat{A})$$

$$= \sum_{x \in S} A^*(x, q_j^*) \qquad (\text{Each } x \in S \text{ is in exactly one } F(t).)$$

$$\leq L \qquad (\text{due to capacity constraint in } C^*)$$

Now, assume that the element $t \in T$ has the maximum distance from its covering center $(\hat{q}_j)$ in $\hat{C}$, as shown in Fig. 2. Since $\hat{A}(t, \hat{q}_j) > 0$, there exists an element $x \in F(t)$ with $A^*(x, q_j^*) > 0$. Let x and q_j^* be members of S_i and $S_{i'}$ for some $i, i' \in \{1, \ldots, m\}$, respectively. So, we can say:

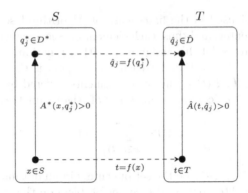

Fig. 2. Element $t \in T$ with the maximum distance from its covering center ($\hat{q}_j$).

$$
\begin{aligned}
R_{T,w'}(\hat{C}) &= d(t, \hat{q}_j) && \text{(by definition of } R_{T,w'}) \\
&\leq d(t, x) + d(x, q_j^*) + d(q_j^*, \hat{q}_j) && \text{(by triangle inequality)} \\
&\leq R_{S_i,w}(C_i) + R_{S,w}(C^*) + R_{S_{i'},w}(C_{i'}) && \text{(by definition of } R_{S_i,w} \text{ and } R_{S,w}) \\
&\leq 2\alpha R_{S,w}(C^*) + R_{S,w}(C^*) + 2\alpha R_{S,w}(C^*) && \text{(by Lemma 1)} \\
&= (4\alpha + 1)R_{S,w}(C^*) && (6)
\end{aligned}
$$

Assume that C^+ is an optimal solution for the weighted balanced k-center problem on input $\langle T, w' \rangle$. Since $\hat{C}$ is a feasible solution for this problem, we have:

$$R_{T,w'}(C^+) \leq R_{T,w'}(\hat{C}) \tag{7}$$

Therefore:

$$
\begin{aligned}
R_{T,w'}(C) &\leq \beta R_{T,w'}(C^+) && \text{(by approximation factor of alg. } \mathcal{B}) \\
&\leq \beta R_{T,w'}(\hat{C}) && \text{(by inequality (7))} \\
&\leq \beta(4\alpha + 1)R_{S,w}(C^*) && \text{(by inequality (6))}
\end{aligned}
$$

$\square$

Theorem 1. *The approximation factor of Algorithm 1 is at most $2\alpha + \beta(4\alpha+1)$ where α and β are the approximation factor of algorithms $\mathcal{A}$ and $\mathcal{B}$.*

Proof. We want to prove that:

$$R_{S,w}(C') \leq (2\alpha + \beta(4\alpha + 1))R_{S,w}(C^*)$$

We have to show that for each element $x \in S$ and each center $c \in D$ covering x ($A'(x, c) > 0$):

$$d(x, c) \leq (2\alpha + \beta(4\alpha + 1))R_{S,w}(C^*)$$

Let x be in S_i (for some $i \in \{1, \ldots, m\}$). Since x is covered by c ($A'(x, c) > 0$), the way of constructing A' implies that there exists at least one intermediate center y in D_i that covers x $(A_i(x,y) > 0)$ and is covered by c $(A(y,c) > 0)$. This is clarified as an example in Fig. 3. A directed edge from a to b here shows that a is covered by b. Note that the separate presentation of sets S_i, D_i, and D in the figure is to give more intuition; we know that D_i is a subset of S_i and might have intersection with D.

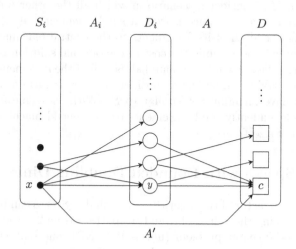

Fig. 3. An example of the covering state of elements in the algorithm

$$d(x,c) \le d(x,y) + d(y,c) \qquad \text{(by triangle inequality)}$$
$$\le R_{S_i,w}(C_i) + R_{T,w'}(C) \qquad \text{(by definition of } R_{S_i,w} \text{ and } R_{T,w'})$$
$$\le 2\alpha R_{S,w}(C^*) + \beta(4\alpha + 1)R_{S,w}(C^*) \quad \text{(by Lemmas 1 and 2)}$$
$$= (2\alpha + \beta(4\alpha + 1))R_{S,w}(C^*)$$

$\square$

Corollary 1. *We have a distributed algorithm for the balanced k-center problem with approximation factor* 49.

Proof. We can achieve the 49-approximation by using Algorithm 1 with proper placement of algorithms $\mathcal{A}$ and $\mathcal{B}$. We can use Gonzalez's algorithm [14] as algorithm $\mathcal{A}$, which yields $\alpha = 2$. For algorithm $\mathcal{B}$, we use the centralized 5-approximation algorithm proposed by Khuller and Sussmann [18] which yields $\beta = 5$. Now, by Theorem 1, the approximation factor of the whole algorithm is $2\alpha + \beta(4\alpha + 1) = 49$. $\square$

The communication complexity of our algorithm is $O(mk)$. We show in the next theorem that this complexity is optimal, under the composable coreset framework.

Theorem 2. *Any distributed algorithm for the capacitated/uncapacitated k-center problem with a bounded approximation guarantee under the composable coreset framework requires $\Omega(mk)$ communication, where k is the number of centers and m is the number of data partitions.*

Proof. Let S_i (the set of elements in the i-th machine) be a set of k points of pairwise distance F. Consider a scenario in which all the other points in other machines are at a small distance ε from a single element $e \in S_i$. If the i-th machine does not send all of its k elements to the central machine, the cost of the final solution will be F while the cost of the optimal solution is ε and thus, the approximation factor will be unbounded. So, all of the k elements in S_i must be sent to the central machine. Now, consider another scenario in which all the local machines have an input set similar to S_i. With the same argument, all of them will independently send k elements to the central machine resulting a communication of size $\Omega(mk)$.   □

4 Metric Spaces with Bounded Doubling Dimension

The approximation factor of the algorithm presented in Sect. 3 can be potentially improved by reducing the values of α and β. As there is no $2 - \varepsilon$ approximation algorithm for the k-center problem (unless $P = NP$), one may conclude that the effect of factor α cannot be reduced to less than 2. However, we show that for special metric spaces, increasing the number of centers picked from each partition S_i can reduce the effect of α. To clarify this, we first introduce a new parameter of metric spaces, which we call "half-coverage constant".

Definition 2. *Let $\mathrm{OPT}(S, k)$ be the cost of an optimal solution for the k-center problem on an input set S. The half-coverage constant of a metric space is the minimum constant c such that for any instance $\langle S, k \rangle$ of the k-center problem in that space,*

$$\mathrm{OPT}(S, ck) \leq \frac{1}{2}\mathrm{OPT}(S, k).$$

According to the above definition, for any arbitrary instance of the k-center problem in a metric space with half-coverage constant H, if the number of centers is multiplied by H, the cost of the optimal solution is reduced by a factor of $1/2$. The following observation is immediate by our definition of half-coverage constant.

Observation 3. *In a metric space with half-coverage constant H, for any input set S and any nonnegative integer t,*

$$\mathrm{OPT}(S, H^t k) \leq \frac{1}{2^t}\mathrm{OPT}(S, k).$$

Not all metric spaces have a bounded half-coverage constant. For example, consider a metric space where the distance of every two distinct elements is 1. If this space has a half-coverage constant H, then for an input set S of size $H + 1$, we have $\text{OPT}(S, 1) = \text{OPT}(S, H) = 1$, which contradicts the definition of half-coverage constant.

Although not all metric spaces have a bounded half-coverage constant, the following theorem shows that every metric space with a bounded doubling dimension (including all $\mathbb{R}^d$ spaces, under any ℓ_p metric) has a bounded half-coverage constant.

Theorem 4. *Every metric space with a doubling constant M has a half-coverage constant at most M^2.*

Proof. Recall that the doubling constant M of a metric space denotes the minimum number of balls of radius $r/2$ needed to cover a ball of radius r. Let *quadrupling constant* Q be the minimum number of balls of radius $r/4$ to cover a ball of radius r. Obviously, $Q \leq M^2$.

Let C^* be an optimal solution for an arbitrary instance $\langle S, k \rangle$ of the k-center problem. For each center $c \in C^*$, consider its coverage ball of radius $\text{OPT}(S, k)$. The ball can be covered by Q balls of radius $\text{OPT}(S, k)/4$, so we can partition the coverage area of c to Q regions in which the distance between any pair of elements is at most $\text{OPT}(S, k)/2$. For each region R, if $S \cap R$ is not empty, select an arbitrary member as a center covering all the elements in $S \cap R$. This results in a feasible solution with at most Qk centers having cost at most $\text{OPT}(S, k)/2$, and thus $\text{OPT}(S, Qk) \leq \text{OPT}(S, k)/2$. So, the half-coverage constant is at most $Q \leq M^2$. $\square$

Now, we show how a bounded half-coverage constant in a metric space can help us achieve better approximation factors for the balanced k-center problem.

Lemma 3. *In a metric space with half-coverage constant H, if we modify the first step of Algorithm 1 to select $H^t k$ (instead of k) centers in each machine, then the approximation factor of Algorithm 1 is reduced to $\beta + \frac{2\alpha + 4\alpha\beta}{2^t}$.*

Proof. The proof is quite similar to Theorem 1. Assume that $C^* = (D^*, A^*)$ is an optimal solution for the whole input set $S = \bigcup_{i=1}^{m} S_i$. We first update the result of Lemma 1 based on the modified algorithm. Define $C^\dagger$, $C_i^\dagger$, and $\hat{C}$ in the same way as in the proof of Lemma 1. In addition, let $C_i^\ddagger$ be an optimal solution for the uncapacitated k-center problem on input S_i with $H^t k$ centers. Now,

$$
\begin{aligned}
R_{S_i, w}(C_i) &\leq \alpha R_{S_i, w}(C_i^\ddagger) && \text{(by approximation factor of algorithm } \mathcal{A}) \\
&\leq \frac{\alpha}{2^t} R_{S_i, w}(C_i^\dagger) && \text{(by Observation 3)} \\
&\leq \frac{\alpha}{2^t} R_{S_i, w}(\hat{C}) && \text{(by inequality (3))} \\
&\leq \frac{2\alpha}{2^t} R_{S, w}(C^\dagger) && \text{(by inequality (4))} \\
&\leq \frac{2\alpha}{2^t} R_{S, w}(C^*) && \text{(by inequality (1))} \qquad (8)
\end{aligned}
$$

We also show that for the modified version of Algorithm 1, we have a result similar to Lemma 2. We can have the same definitions of $\hat{C}$ and C^+ as its proof and will then have:

$$
\begin{aligned}
R_{T,w'}(C) &\leq \beta R_{T,w'}(C^+) && \text{(by algorithm } \mathcal{B}) \\
&\leq \beta R_{T,w'}(\hat{C}) && \text{(by (7))} \\
&\leq \beta(R_{S_i,w}(C_i) + R_{S,w}(C^*) + R_{S_{i'},w}(C_{i'})) && \text{(similar to (6))} \\
&\leq \beta(\frac{2\alpha}{2^t} R_{S,w}(C^*) + R_{S,w}(C^*) + \frac{2\alpha}{2^t} R_{S,w}(C^*)) && \text{(by (8))} \\
&= \beta(1 + \frac{4\alpha}{2^t}) R_{S,w}(C^*) && (9)
\end{aligned}
$$

We finally prove this theorem similar to the proof of Theorem 1. Let $x \in S_i$ (for some $i \in \{1, \ldots, m\}$) have the maximum distance to its covering center c in C'. Therefore,

$$
\begin{aligned}
R_{S,w}(C') &= d(x, c) && \text{(by definition of } R_{S,w}) \\
&\leq R_{S_i,w}(C_i) + R_{T,w'}(C) && \text{(similar to proof of Theorem 1)} \\
&\leq \frac{2\alpha}{2^t} R_{S,w}(C^*) + \beta(1 + \frac{4\alpha}{2^t}) R_{S,w}(C^*) && \text{(by inequalities (8) and (9))} \\
&= (\beta + \frac{2\alpha + 4\alpha\beta}{2^t}) R_{S,w}(C^*),
\end{aligned}
$$

which completes the proof.   □

Theorem 5. *There is a distributed algorithm with an approximation factor of $5 + \varepsilon$ for the weighted balanced k-center problem in metric spaces with bounded doubling dimension.*

Proof. Again, we use Gonzalez's algorithm [14] for algorithm $\mathcal{A}$, and the algorithm of Khuller and Sussmann [18] for algorithm $\mathcal{B}$, yielding $\alpha = 2$ and $\beta = 5$. Therefore, by Lemma 3, the approximation factor of our algorithm is

$$
\beta + \frac{2\alpha + 4\alpha\beta}{2^t} = 5 + \frac{44}{2^t} \leq 5 + \varepsilon,
$$

which holds for $t = \lceil \log_2(\frac{44}{\varepsilon}) \rceil$. Furthermore, the communication complexity of the algorithm, i.e. $O(mk)$, is multiplied by $H^t \approx (\frac{44}{\varepsilon})^{\log_2 H}$, which is poly($\frac{1}{\varepsilon}$). □

5   Conclusion

In this paper, we presented a new approximation algorithm for the weighted balanced k-center problem, improving over the best current algorithm of Bateni et al. [4]. We also showed that the approximation factor of our algorithm can be improved to $5 + \varepsilon$ in some metric spaces including those with bounded doubling dimension. The ideas used in this paper seems applicable to other variants of the centroid-based clustering, including k-median and k-means, and hence are open for further investigation.

References

1. Aghamolaei, S., Farhadi, M., Zarrabi-Zadeh, H.: Diversity maximization via composable coresets. In: Proceedings of 27th Canadian Conference on Computational Geometry, p. 43 (2015)
2. An, H.-C., Bhaskara, A., Chekuri, C., Gupta, S., Madan, V., Svensson, O.: Centrality of trees for capacitated k-center. Math. Program. **154**(1), 29–53 (2015). https://doi.org/10.1007/s10107-014-0857-y
3. Barilan, J., Kortsarz, G., Peleg, D.: How to allocate network centers. J. Algorithms **15**(3), 385–415 (1993)
4. Bateni, M., Bhaskara, A., Lattanzi, S., Mirrokni, V.: Distributed balanced clustering via mapping coresets. In: Proceedings of the 27th Annual Conference on Neural Information Processing Systems, pp. 2591–2599 (2014)
5. Ceccarello, M., Pietracaprina, A., Pucci, G.: Solving k-center clustering (with outliers) in mapreduce and streaming, almost as accurately as sequentially. In: Proceedings of the 45th International Conference on Very Large Data Bases, vol. 12, pp. 766–778 (2019)
6. Chakrabarty, D., Krishnaswamy, R., Kumar, A.: The heterogeneous capacitated k-center problem. In: Eisenbrand, F., Koenemann, J. (eds.) IPCO 2017. LNCS, vol. 10328, pp. 123–135. Springer, Cham (2017). https://doi.org/10.1007/978-3-319-59250-3_11
7. Chen, K.: On coresets for k-median and k-means clustering in metric and Euclidean spaces and their applications. SIAM J. Comput. **39**(3), 923–947 (2009)
8. Cygan, M., Hajiaghayi, M., Khuller, S.: LP rounding for k-centers with non-uniform hard capacities. In: Proceedings of the 53rd Annual IEEE Symposium on Foundations of Computer Science, pp. 273–282 (2012)
9. Cygan, M., Kociumaka, T.: Constant factor approximation for capacitated k-center with outliers. In: Proceedings of the 31st Symposium on Theoretical Aspects of Computer Science, STACS 2014, vol. 25, pp. 251–262 (2014)
10. Ding, H.: Balanced k-center clustering when k is a constant. In: Proceedings of the 29th Canadian Conference on Computational Geometry, pp. 179–184 (2017)
11. Ding, H., Hu, L., Huang, L., Li, J.: Capacitated center problems with two-sided bounds and outliers. WADS 2017. LNCS, vol. 10389, pp. 325–336. Springer, Cham (2017). https://doi.org/10.1007/978-3-319-62127-2_28
12. Ene, A., Im, S., Moseley, B.: Fast clustering using mapreduce. In: Proceedings of the 17th ACM SIGKDD International Conference on Knowledge Discovery and Data Mining, pp. 681–689 (2011)
13. Fernandes, C.G., de Paula, S.P., Pedrosa, L.L.C.: Improved approximation algorithms for capacitated fault-tolerant k-center. In: Proceedings of the12th Latin American Theoretical Information on Symposium, pp. 441–453 (2016)
14. Gonzalez, T.F.: Clustering to minimize the maximum intercluster distance. Theoret. Comput. Sci. **38**, 293–306 (1985)
15. Hochbaum, D.S., Shmoys, D.B.: A best possible heuristic for the k-center problem. Math. Oper. Res. **10**(2), 180–184 (1985)
16. Im, S., Moseley, B.: Brief announcement: fast and better distributed mapreduce algorithms for k-center clustering. In: Proceedings of the 27th ACM Symposium on Parallel Algorithms Architecture, pp. 65–67 (2015)
17. Indyk, P., Mahabadi, S., Mahdian, M., Mirrokni, V.S.: Composable core-sets for diversity and coverage maximization. In: Proceedings of the 33rd Symposium on Principles of Database Systems, pp. 100–108 (2014)

18. Khuller, S., Sussmann, Y.J.: The capacitated k-center problem. SIAM J. Discrete Math. **13**(3), 403–418 (2000)
19. Malkomes, G., Kusner, M.J., Chen, W., Weinberger, K.Q., Moseley, B.: Fast distributed k-center clustering with outliers on massive data. In: Proceedings of the 28th Annual Conference on Neural Information Processing Systems, pp. 1063–1071 (2015)
20. McClintock, J., Wirth, A.: Efficient parallel algorithms for k-center clustering. In: Proceedings of the 45th International Conference on Parallel Processing, pp. 133–138 (2016)
21. Matthew McCutchen, R., Khuller, S.: Streaming algorithms for k-center clustering with outliers and with anonymity. In: Goel, A., Jansen, K., Rolim, J.D.P., Rubinfeld, R. (eds.) APPROX/RANDOM-2008. LNCS, vol. 5171, pp. 165–178. Springer, Heidelberg (2008). https://doi.org/10.1007/978-3-540-85363-3_14
22. Mirjalali, K., Tabatabaee, S.A., Zarrabi-Zadeh, H.: Distributed unit clustering. In: Proceedings of the 31st Canadian Conference on Computational Geometry, pp. 236–241 (2019)
23. Mirrokni, V.S., Zadimoghaddam, M.: Randomized composable core-sets for distributed submodular maximization. In: Proceedings of the 47th Annual ACM Symposium on Theory of Computing, pp. 153–162 (2015)

Finite Interval-Time Transition System for Real-Time Actors

Shaghayegh Tavassoli, Ramtin Khosravi$^{(\boxtimes)}$, and Ehsan Khamespanah

School of Electrical and Computer Engineering, University of Tehran, Tehran, Iran
r.khosravi@ut.ac.ir

Abstract. Real-time computer systems are software or hardware systems which have to perform their tasks according to a time schedule. Formal verification is a widely used technique to make sure if a real-time system has correct time behavior. Using formal methods requires providing support for non-deterministic specification for time constraints which is realized by time intervals. Timed-Rebeca is an actor-based modeling language which is equipped with a verification tool. The values of timing features in this language are positive integer numbers and zero (discrete values). In this paper, Timed-Rebeca is extended to support modeling timed actor systems with time intervals. The semantics of this extension is defined in terms of Interval-Time Transition System (ITTS) which is developed based on the standard semantics of Timed-Rebeca. In ITTS, instead of integer values, time intervals are associated with system states and the notion of shift equivalence relation in ITTS is used to make the transition system finite. As there is a bisimulation relation between the states of ITTS and finite ITTS, it can be used for verification against branching-time properties.

Keywords: Actor model · Timed Rebeca · Interval-Time Transition System · Bisimulation relation

1 Introduction

Real-time computer systems, are hardware or software systems which work on the basis of a time schedule [15]. Controller of car engines, networks of wireless sensors and actuators, and multimedia data streaming applications are examples of real-time systems. The correct behavior of real-time systems is achieved by correctness of calculated values as well as the time those values were produced [14,22]. So, verification of a real-time system requires considering both the functional and time behavior of the systems [15]. In the modeling of real-time systems, presenting nondeterministic time behavior may be required. For example, the best response time of drivers when braking (with a high probability of signal prediction) was reported as a value in the range of 0.7 to 0.75 s [8]. Using time intervals is a widely used notion to model such behavior. Supporting time intervals for nondeterministic time behavior raises challenges in the schedulability analysis of real-time systems [16].

© IFIP International Federation for Information Processing 2020
Published by Springer Nature Switzerland AG 2020
L. S. Barbosa and M. Ali Abam (Eds.): TTCS 2020, LNCS 12281, pp. 85–100, 2020.
https://doi.org/10.1007/978-3-030-57852-7_7

Examples of nondeterministic time behavior in real-time systems include the network delay in communication systems, driver's reaction time when braking a car, and the execution time of programs on processors. In such cases, determining the exact duration of the tasks is not always possible. Instead, a time interval can be used to specify that the time value will be within a specific range. It is well-known that Worst Case Execution Time (WCET) analysis does not necessarily show the worst time behavior [7]. Hence, using the upper bounds for the actions does not help when analyzing time-sensitive safety properties. In Timed Rebeca, the values of timing primitives are discrete values [12,13]. Therefore, it is useful to propose an extension on Timed Rebeca for modeling and analyzing real-time systems with time intervals.

Using formal methods, in general, and model checking [6] in particular, is a verification technique which guarantees correctness of systems. Model checking tools for real-time models exhaustively explore state spaces of systems to make sure that given properties hold in all possible executions of the system and specified time constraints are satisfied. *Timed Automata* and *UPPAAL* are widely used for modeling and model checking of real-time systems [4]. Timed Automata are automata in which clock constraints (i.e. time interval constraints associated with clocks) can be associated with both transition guards and location invariants [3].

In the context of distributed systems, *Actor* is used to model systems composed of a number of distributed components communicating via message passing [2,10]. There are some extensions on the Actor model for modeling real-time systems, e.g. *Timed Rebeca* [17] and *Creol* [5], which provide model checking for timed properties such as schedulability and deadlock freedom. Creol models are transformed to timed automata for the model checking purpose which suffers from support for simple expressions for time constrains and state space explosion for middle-sized models. In contrast, Timed Rebeca provides direct model checking toolset; but, only for models with discrete timing primitives [11,12] as described in detail in Sect. 2.

In this paper, we extend Timed Rebeca to support modeling and analysis of real-time systems with time intervals (Sect. 3). To this end, the notion of Interval-Time Transition System (ITTS) is introduced, which can serve as the basis for the semantic description of timed actor systems. Here, we formally describe the semantics of Timed Rebeca based on ITTS in Sect. 4. In the second step, we illustrate how the notion of *shift equivalence relation* for ITTS is used to make the transition system finite, if possible. Using bisimulation method, in Sect. 5, it is proved that shift equivalence relation can be used for detecting equivalent system states in ITTS.

2   Timed Actors

A well-established paradigm for modeling concurrent and distributed systems is the *Actor* model. Actor is introduced by Hewitt [10] as an agent-based language and is later developed as a model of concurrent computation by Agha [2].

Actors are seen as the universal primitives of concurrent computation, such that each actor provides some services which can be requested by other actors by sending messages to it. Execution of a service of an actor may result in changing the state of the actor and sending messages to some other actors. In [1], Timed Rebeca is introduced as an extension on the actor model for modeling of real-time systems.

2.1   Timed Rebeca

Timed Rebeca is introduced in [1] as the real-time extension of the Rebeca modeling language [18, 19, 21]. We explain Timed Rebeca using the example of a simple Ping-Pong model (taken from [20] with slight modifications). In this model, the ping actor sends *pong* message to the pong actor and the pong actor sends *ping* message to the ping actor. A Timed Rebeca model consists of a number of *reactive classes*, each describing the type of a number of *actors* (also called *rebecs*)[1]. There are two reactive classes PingClass and PongClass in the Ping-Pong model (Listing 1 lines 1 and 16). Each reactive class declares a set of *state variables* (e.g. lines 5–7). The local state of each actor is defined by the values of its state variables and the contents of its message bag. Communication in Timed Rebeca models takes place by asynchronous message passing among actors. Each actor has a set of *known rebecs* to which it can send messages. For example, an actor of type PingClass knows an actor of type PongClass (line 3), to which it can send pong message (line 12). Reactive classes may have some constructors that have the same name as the declaring reactive class and do not have a return value (lines 8–10 for PingClass). They may initialize actor's state variables and put the initially needed messages in the *message bag* of that actor (line 9). The way of responds to a message is specified in a *message server* which are methods defined in reactive classes. An actor can change its state variables through assignment statements, makes decisions through conditional statements, communicates with other actors by sending messages (e.g., line 12), and performs periodic behavior by sending messages to itself. Since communication is asynchronous, each actor has a *message bag* from which it takes the next incoming message. The ordering of the messages in a message bag is based on the arrival times of messages. An actor takes the first message from its message bag, executes its corresponding message server in an isolated environment, takes the next message (or waits for the next message to arrive) and so on.

Finally, the **main** block is used to instantiate the actors of the model. In the Ping-Pong model, two actors are created receiving their known rebecs and the parameter values of their constructors upon instantiation (lines 26–27).

```
1   reactiveclass PingClass(3) {          7       }
2       knownrebecs {                     8       PingClass() {
3           PongClass pong1;              9           self.ping();
4       }                                10       }
5       statevars {                      11       msgsrv ping() {
6           //e.g. int var1, var2;       12           pong1.pong() after(1);
```

---

[1] In this paper we use rebec and actor interchangeably.

```
13          delay(2);                    23        }
14        }                              24    }
15  }                                    25  main {
16  reactiveclass PongClass(3) {         26        PingClass pi(po) : ();
17      knownrebecs {                    27        PongClass po(pi) : ();
18          PingClass ping1;             28  }
19      }
20      msgsrv pong() {
21          ping1.ping() after(1);
22          delay(1);
```

Listing. 1. The Timed Rebeca model of the Ping-Pong model

Timed Rebeca adds three primitives to Rebeca to address timing issues: *delay, deadline* and *after*. A *delay* statement models the passage of time for an actor during execution of a message server (line 13). Note that all other statements of Timed Rebeca are assumed to execute instantaneously. The keywords *after* and *deadline* are used in conjunction with a method call. The term after(n) indicates that it takes n units of time for a message to be delivered to its receiver (lines 12 and 21). The term deadline(n) expresses that if the message is not taken in n units of time, it will be purged from the receiver's message bag automatically [1].

2.2 Timed Transition Systems

One way of modeling real-time systems is using timed transition systems [9]. The Semantics of Timed Rebeca is defined in terms of timed transition systems in [11]. In this semantics, the global state of a Timed Rebeca model is represented by a function that maps actors' ids to tuples. A tuple contains, the state variables valuations, the content of message bags, local time, the program-counter which shows the position of the statements which will be executed to finish the service to the message currently being processed, and the time when the actor resumes execution of the remained statements.

Transitions between states occur as the results of actors' activities including: taking a message from the mailbox, continuing the execution of statements, and progress of time. In timed transition system of Timed Rebeca, progress of time is only enabled when the other actions are disabled for all of the actors. This rule performs the minimum required progress of time to make one of the other rules enabled. As a result, the model of progress of time in the timed transition system of Timed Rebeca is deterministic. The detailed SOS rules of transition relations are defined in [11].

Figure 1 shows the beginning part of the timed transition system of the Ping-Pong example. In this figure, σ denotes the next statement in the body of the message server which is being processed. The pair of $\sigma = \langle ping, 1 \rangle$ in the second state of Fig. 1 shows that pi executed the statements of its ping message server up to the statement in line 1. In each state, r is the time when the actor can resume the execution of its remained statements. The first enabled actor of the model is pi (as its corresponding reactive class PingClass has a constructor which puts message ping in its bag, line 9), so, the first possible transition is taking message ping. As shown in the detailed contents of the second state (the gray block), taking the message ping results in setting the values of σ and r for the actor pi. The next transition results in executing the first statement of the message server ping and results in putting the message pong in the bag of the actor po with release time 1 (because of the value of after in line 7). Note that the deadline for messages in this model is ∞ as no specific value is set as the deadline for these messages. As the next statement of the message server ping is a delay statement, pi cannot continue the execution. The actor po cannot cause a transition too. So, the only possible transition is progress of time which is by 1 unit from the third to the fourth state.

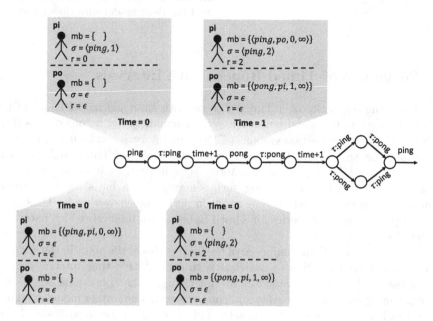

Fig. 1. The beginning part of the transition system of the Ping-Pong example [20].

3 Timed Rebeca with Intervals

The time interval extension to Timed Rebeca enables the use of time intervals with the timing directive after. In this model, a time interval is associated with each state of the transition system. To simplify the presentation of this paper, the time features delay and deadline are omitted. The modified version of Listing 1 which contains time interval for after directive is presented in Listing 2. Note that in Line 9, after([4,8]) means that message pong arrives its destination during [4, 8) time units after it has been sent. This value models the nondeterministic delay of the network in message delivery.

```
1  reactiveclass PingClass(3) {        16      PongClass() {
2      knownrebecs {                   17          self.pong();
3          PongClass po;               18      }
4      }                               19      msgsrv pong() {
5      PingClass() {                   20          pi.ping() after([4,8));
6          self.ping();                21      }
7      }                               22  }
8      msgsrv ping() {                 23  main {
9          po.pong() after([4,8));     24      PingClass pi(po) : ();
10     }                               25      PongClass po(pi) : ();
11 }                                   26  }
12 reactiveclass PongClass(3) {
13     knownrebecs {
14         PingClass pi;
15     }
```

Listing. 2. The Timed Rebeca model of the Ping-Pong model with time intervals

4   Semantics of Timed Rebeca with Intervals

We define the semantics of a Timed Rebeca with Intervals model as an ITTS which is based on the usual transition system semantics for actor systems [11]. Although in timed transition systems (TTS) time intervals can be used as timing constraints on transitions [9], describing the semantics of Timed Rebeca with intervals using TTS makes the semantics complicated, as it does not fit the timing model of TTS. Also, using such semantics as a basis for state space generation and model analysis may result in performance overheads. Thus, it is necessary to define a semantics for Timed Rebeca with intervals which naturally reflects its timing model.

The states in ITTS are composed of the local states of the actor in the system. A key idea behind ITTS is to associate with each state a time interval during which the local states of the actors do not change. The transitions are of two types, namely *message processing* and *time progress*. The former includes taking or processing of a message by an actor (which changes the local state of the actor). Time progress transition changes the state of the system by increasing the left end-point of the time interval of the state.

To keep the semantics description focused on timed behavior, we assume the messages do not have parameters, as they do not affect the time behavior of the model. Otherwise, the semantics rules will be cluttered with actual parameters evaluation and scope management.

Before we describe the formal semantics of ITTS, we introduce a few notations that are used throughout the paper.

4.1   Notation and Basic Definitions

We use the notation *TInterval* (ranged over by α, β, and γ) to denote the set of all time intervals in $R_{\geq 0}$ which are left-closed right-open or left-closed right-closed. Such intervals are written as $[t_1, t_2)$ and $[t_1, t_2]$ respectively. Time interval $[t_1, t_1]$ corresponds to time value t_1 of Timed Rebeca. For $\alpha \in TInterval$, α_ℓ and α_r denotes the left and the right endpoints of α respectively (regardless of being right-open or right-closed). The notation $\alpha_{\ell \leftarrow x}$ (resp. $\alpha_{r \leftarrow x}$) denote the interval obtained from replacing the left (resp. right) endpoint of α with x, e.g., $[t_1, t_2)_{r \leftarrow x}$ is $[t_1, x)$.

For a function $f : X \rightarrow Y$, we use the notation $f[x \mapsto y]$ to denote the function $\{(a, b) \in f | a \neq x\} \cup \{(x, y)\}$. We also use the notation $x \mapsto y$ as an alternative to (x, y).

The following notations is used for working with sequences. Given a set A, the set A^* is the set of all finite sequences over elements of A, ranged over by σ and σ'. For a sequence $\sigma \in A^*$ of length n, the symbol a_i denotes the ith element of the sequence, where $1 \leq i \leq n$. In this case, we may also write σ as $\langle a_1, a_2, \ldots, a_n \rangle$. The empty sequence is represented by ϵ, and $\langle h | T \rangle$ denotes a sequence whose first element is $h \in A$ and $T \in A^*$ is the sequence comprising the elements in the rest of the sequence. For $x \in A$ and $\sigma, \sigma' \in A^*$, the expressions $x \oplus \sigma$ and $\sigma \oplus x$ denote a sequence obtained from prepending and appending x to σ respectively. Also, $\sigma \oplus \sigma'$ is the sequence obtained by concatenating σ' to the end of σ. The *deletion* operator is defined such that $\sigma \ominus x$ is the sequence obtained by removing the first occurrence of x in σ. Finally, we define the membership operator for sequences as $x \in \langle a_1, a_2, \ldots, a_n \rangle \overset{def}{=} \exists 1 \leq i \leq n \cdot a_i = x$.

VName is defined as the set of all variable names in the model, and *Value* as the set of all values the state variables can take. We do not address typing issues in this paper to keep the focus on time behavior of the actors. Let *Stat* denote the set of all statements appearing in the actors' message handlers, *MName* denote the set of all message names, and *ActorID* denote the set of all unique identities of the actors. The function $body : ActorID \times MName \rightarrow Stat$ is defined such that $body(x, m)$ returns the sequence of statements in the body of the message handler of m in the actor with ID x. The set *Act* is defined as $MName \cup \{\tau\} \cup \{TP\}$.

4.2   Messages and Message Bags

A message in ITTS is a tuple (sID, rID, m, α), where rID and sID denote the receiver ID and the sender ID respectively, m denotes the message name, while α denotes the "after" interval. It means that this message arrives its destination during α time units after it has been sent. The set of all messages is defined as *Msg*. In ITTS, each actor has an associated message bag that holds all its received messages. The set of all message bags is defined as $Bag(Msg) = Msg^*$. For the notation to be simpler, message bags are written in the form of sequences. But it is not necessary for message bags to be sequences, as they can be defined as multisets in general.

4.3   States in ITTS

The system state is composed of the local states of the actors in the system. The local state of an actor with ID x, is defined as the triple (v_x, mb_x, σ_x), where v_x is the valuation function of the state variables of the actor, mb_x is the message bag of the actor, and σ_x denotes the sequence of statements the actor is going to execute in order to finish the processing of the message being handled.

The set of all local states of the actors is denoted as $ActorState = (VName \rightarrow Value) \times Bag(Msg) \times Stat^*$.

A (global) system state in ITTS is a tuple (s, α), where $s \in ActorID \rightarrow ActorState$ is a function mapping each actor ID to its local state, and $\alpha \in TInterval$ is the time interval associated with the system state. It is assumed that S is the set of all possible system states, ranged over by gs (short for *global state*). In the initial system state, we let the time interval of the system state to be $[0, 0]$.

4.4   Order of Events in ITTS

As stated before, to define *time progress* transitions, we must determine the earliest time in which a change is possible in the local state of an actor (called an *event*). As such changes happen only as a result of taking or executing a message, the earliest and the latest times the messages can be taken determine the order of events in the system. To specify the ordering of the events, a message bag is defined for every system state called "system state message bag", denoted by $B(gs)$, which consists of all messages in message bags of all actors in that system state (which may contain duplicate messages).

The ordering of the events regarding to a state $gs \in S$ is denoted by $EE_i(gs)$ which is the ith smallest value in the set of all lower and upper bounds of the time intervals of all messages in $B(gs)$.

4.5   Transitions Definition

As explained before, transitions in ITTS are classified into two major types: *message processing* and *time progress*. The former includes taking a message from message bag or processing a message by an actor. As with Timed Rebeca, we assume executing the statements in a message server is instantaneous.

Message Processing. For a system state (s, α) we call an actor x *idle*, if it has no remaining statement to execute from its previously taken message. If such an actor has a message in its mailbox whose *after* interval starts from α_ℓ, a message processing transition can take place:

$$\frac{gs = (s, \alpha) \in S \land s(x) = (v, mb, \epsilon) \land msg = (y, x, m, \beta) \in mb \land \beta_\ell = \alpha_\ell}{gs \xrightarrow{m} (s[x \mapsto (v \cup \{(self, x), (sender, y)\}, mb \ominus msg, body(x, msg))], \alpha)} \quad (1)$$

The execution of each statement in the body of a message handler is considered an internal action in ITTS. The statements such as assignments, conditionals and loops only alter the local state of the executing actor. The only statement that affects the state of other actors is *send* which may put a message in another actor's message bag. The semantics of send and assignment statements are stated here and the others are left out to save space.

An assignment statement of the form $var = expr$ overrides the value of var in the executing actor's state variables to the value of the expression $expr$, denoted by $eval_x(expr)$. To keep the description simple, it is assumed that the message servers do not have local variables, so the left side of the assignment is always a state variable of the actor executing the message server.

$$\frac{gs = (s, \alpha) \in S \land s(x) = (v, mb, \langle var = expr | \sigma \rangle)}{gs \rightarrow (s[x \mapsto (v[var \mapsto eval_x(expr)], mb, \sigma)], \alpha)} \quad (2)$$

The send statement $y.m()after\ \gamma$ denotes sending a message m with the after interval γ to the receiver y. The semantics of the send statement is defined as

$$\frac{gs = (s, \alpha) \in S \land s(x) = (v, mb, \langle y.m()after\ \gamma | \sigma \rangle) \land s(y) = (v', mb', \sigma')}{gs \rightarrow (s[x \mapsto (v, mb, \sigma)][y \mapsto (v', mb' \oplus (x, y, m, \beta), \sigma')], \alpha_{r \leftarrow t})} \quad (3)$$

where β is the interval whose left and right endpoints are $\alpha_\ell + \gamma_\ell$ and $\alpha_r + \gamma_r$ respectively, and is right-closed if both α and γ are right-closed and is right-open otherwise. Furthermore, $t = min(\alpha_r, \beta_\ell)$ which means that after the message is sent, β_ℓ may be the second earliest event, replacing α_r in that case.

Time Progress. In ITTS, two types of time progress transitions are defined. The first type corresponds to the case that the only possible transition is time progress, while for the second type, a nondeterministic choice between time progress and another transition is enabled.

Type 1 Time Progress Transition: If the lower bound of the time interval for a system state gs is smaller than the earliest event of that system state ($\alpha_\ell < EE_1(gs)$), the only possible transition is time progress. After executing Type 1 time progress transition, the time interval for the successor state is $[EE_1(gs), EE_2(gs))$.

$$\frac{gs = (s, \alpha) \land \alpha_\ell < EE_1(gs)}{gs \xrightarrow{TP} (s, [EE_1(gs), EE_2(gs))} \quad (4)$$

Type 2 Time Progress Transition: In ITTS, in situations which in a system state, time progress transition and at least one other transition are enabled, second type of time progress transition can occur. Consider a global state $gs = (s, \alpha)$. Unlike Type 1 transitions we have $\alpha_\ell = EE_1(gs)$, so there exists a message msg_1 in the system with interval β such that $\alpha_\ell = \beta_\ell$. Now, if no other message exists which can be taken before β_r, the only possible transition from gs is to take msg_1 and there is no time progress transition. Hence, there must be a message msg_2 in the system with interval γ such that $EE_2(gs) = \gamma_\ell$ and $\beta_\ell \leq \gamma_\ell < \beta_r$. In such a state gs, two transitions are possible: taking msg_1, and waiting till γ_ℓ. Note that this nondeterminism enables the interleaving of processing msg_1 and msg_2.

To model this type of time progress, we shift the lower bound of the time interval of msg_1 from β_ℓ to γ_ℓ and update the time interval of the global state accordingly. Note that there may be multiple messages that start from β_ℓ. Hence, we define the function ds (short for $delay\ starts$), such that $ds(mb, t)$ changes the lower bound of the messages in mb which start earlier than t to t. Formally,

$$ds(\epsilon, t) = \epsilon$$
$$ds(\langle (x, y, m, \alpha | T \rangle, t) = \langle (x, y, m, \alpha_{\ell \leftarrow max(\alpha_\ell, t)}) | ds(T, t) \rangle$$

We lift the definition of ds to the function s which returns the local state of the actors:

$$ds(s, t) = \{ x \mapsto (v, ds(mb, t), \sigma) | x \mapsto (v, mb, \sigma) \in s \}$$

Now, we can define Type 2 time progress transitions as below.

$$\frac{gs = (s, \alpha) \wedge \alpha_\ell = EE_1(gs) \wedge (x, y, m, \gamma) \in B(gs) \wedge \gamma_\ell = EE_2(gs)}{gs \xrightarrow{TP} (ds(s, EE_2(gs)), [EE_2(gs), EE_3(gs))} \quad (5)$$

Figure 2 shows part of the ITTS of the Ping-Pong example of Listing 2 which is generated based on the proposed semantics of this section.

5   Making State Space Finite

Based on the semantics of Timed Rebeca, there is no explicit time reset operator in the language; so, the progress of time results in an infinite number of states in ITTS of models. However, reactive systems which generally show periodic or recurrent behaviors are modeled using Timed Rebeca, i.e. performing recurrent behaviors over infinite time. This fact enables us to propose the notion for equivalence relation between two states with time intervals, aiming to make ITTSs finite, called *shift equivalence relation in ITTS*. The idea of defining shift equivalence relation in states of ITTSs is inspired from [12]. Intuitively, in the shift equivalence relation two states are equivalent if and only if they are the same except for the parts related to the time and the timed parts can be mapped by shifting them with an specific value.

5.1   Shift Equivalence Relation in ITTS

In the first step, the shift equivalence of two time intervals with distance $c \in R$ is defined as:

$$\alpha \approx_c \alpha' \stackrel{def}{=} \alpha'_\ell = \alpha_\ell + c \wedge \alpha'_r = \alpha_r + c \quad (6)$$

Then, the shift equivalence of two messages with distance $c \in R$ is defined as:

$$\forall sID, rID \in ActorID, m \in MName, \beta, \beta' \in TInterval \cdot$$
$$(sID, rID, m, \beta) \approx_c (sID, rID, m, \beta') \Leftrightarrow \beta \approx_c \beta' \quad (7)$$

Using Eq. 7, the shift equivalence of two message bags $B = \langle m_1, ..., m_n \rangle$ and $B' \langle m'_1, ..., m'_n \rangle$ with distance $c \in R$ is defined as:

$$B \approx_c B' \Leftrightarrow \exists 1 \leq i \leq n, 1 \leq i' \leq n \cdot m_i \approx_c m'_i \wedge B \ominus m_i \approx_c B' \ominus m'_{i'} \quad (8)$$

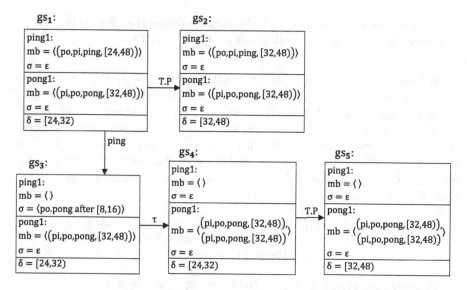

Fig. 2. A part of the interval time transition system of the Ping-Pong example. The transition from gs_1 to gs_3 is a take message transition. The transition from gs_3 to gs_4 is an internal transition. The transition from gs_4 to gs_5 is a Type 1 time progress. The transition from gs_1 to gs_2 is a Type 2 time progress.

Now, the shift equivalence of two local states of an actor with ID x with distance $c \in R$ is defined as:

$$(s(x) = (v_x, mb_x, \sigma_x)) \equiv_c (s'(x) = (v'_x, mb'_x, \sigma'_x)) \Leftrightarrow$$
$$v_x = v'_x \wedge \sigma_x = \sigma'_x \wedge mb_x \approx_c mb'_x \quad (9)$$

Consequently, the shift equivalence of two system states gs and gs' ($gs, gs' \in S$) with distance $c \in R$ is defined as:

$$\big(gs = (s, \alpha)\big) \equiv_c \big(gs' = (s', \alpha')\big) \Leftrightarrow \alpha \approx_c \alpha' \wedge \forall x \in ActorID.s(x) \equiv_c s'(x) \quad (10)$$

5.2   Shift Equivalence Relation in ITTS Is a Bisimulation Relation

The shift equivalence relation aims to make ITTSs of models finite. To this end, we have to show that there is a timed bisimulation relation between finite and infinite ITTSs of a given model to prove that they preserve the same set of timed branching-time properties (i.e., $\equiv_c$ is a bi-simulation relation). To this end, the following theorem should be proven:

Theorem 1. $\forall(gs_1, gs'_1) \in \equiv_c$ and $\forall a \in Act$:

$$\forall gs_2 \in S.gs_1 \xrightarrow{a} gs_2 \Rightarrow \exists gs'_2 \in S.gs'_1 \xrightarrow{a} gs'_2 \wedge (gs_2, gs'_2) \in \equiv_c \quad (11)$$

$$\forall gs'_2 \in S.gs'_1 \xrightarrow{a} gs'_2 \Rightarrow \exists gs_2 \in S.gs_1 \xrightarrow{a} gs_2 \wedge (gs_2, gs'_2) \in \equiv_c \quad (12)$$

To prove the first condition of bisimulation relation (i.e., Eq. (11)) it has to be proven that it holds for for all transition types in ITTS.

Take Message Transitions: Assume that $gs_1 = (s_1, \alpha)$, $s_1(x) = (v_x, mb_x, \epsilon)$, $msg = (y, x, m, \beta) \in mb_x$ and $\beta_\ell = \alpha_\ell$. Thus take message transition is enabled in gs_1 and $gs_1 \overset{m}{\to} gs_2$. Assume that $gs_1' = (s_1', \alpha')$, $s_1'(x) = (v_x', mb_x', \epsilon)$ and $gs_1 \equiv_c gs_1'$. Therefore $\alpha_\ell' = \alpha_\ell + c$. From $s_1(x) \equiv_c s_1'(x)$, it can be concluded that a message msg' exists in mb_x' such that $msg \approx_c msg'$. Thus msg' is of the form (y, x, m, β') such that $\beta_\ell' = \alpha_\ell'$ and $\beta_r' = \beta_r + c$. Therefore take message transition is enabled in gs_1'. Hence, gs_2' exists in ITTS such that $gs_1' \overset{m}{\to} gs_2'$. From Eq. (1), It can be concluded that:

$$gs_2 = (s_2, \alpha) = (s_1[x \mapsto (v_x \cup A, mb_x \ominus msg, body(x, msg)], \alpha)$$
$$gs_2' = (s_2', \alpha') = (s_1'[x \mapsto (v_x' \cup A, mb_x' \ominus msg', body(x, msg')], \alpha')$$
$$A = \{(self, x), (sender, y)\}$$

To prove that $gs_2(x) \equiv_c gs_2'(x)$, the following points should be considered:

1. $s_1(x) \equiv_c s_1'(x) \Rightarrow v_x = v_x'$
   $\Rightarrow v_x \cup \{(self, x), (sender, y)\} = v_x' \cup \{(self, x), (sender, y)\}$
2. $mb_x \approx_c mb_x' \wedge msg \approx_c msg' \Rightarrow mb_x \ominus msg \approx_c mb_x' \ominus msg'$
3. $msg \approx_c msg' \Rightarrow body(x, msg) = body(x, msg')$
4. $gs_1 \equiv_c gs_1' \Rightarrow \alpha \approx_c \alpha'$

Based on the above equations, $gs_2 \equiv_c gs_2'$ can be concluded, which is required for proving Eq. 11.

Internal Transitions: Assuming that $gs_1 = (s_1, \alpha)$, $gs_1' = (s_1', \alpha')$, $gs_1 \equiv_c gs_1'$, two cases are possible:

– **Assignment Statement:** Assume that $s_1(x) = (v_x, mb_{1x}, \langle var = expr | \sigma_x \rangle)$. Therefore internal action is enabled in gs_1 and $gs_1 \overset{\tau}{\to} gs_2$. From $gs_1 \equiv_c gs_1'$, it can be concluded that $s_1(x) \equiv_c s_1'(x)$, so $s_1'(x) = (v_x, mb_{1x}', \langle var = expr | \sigma_x \rangle)$. Thus internal action is enabled in gs_1' and $gs_1' \overset{\tau}{\to} gs_2'$. The produced next states gs_2 and gs_2' are:

$$gs_2 = (s_1[x \mapsto (v_x[var \mapsto eval_x(expr)], mb_{1x}, \sigma_x)], \alpha)$$
$$gs_2' = (s_1'[x \mapsto (v_x[var \mapsto eval_x(expr)], mb_{1x}', \sigma_x)], \alpha')$$

To prove that $gs_2 \equiv_c gs_2'$, the following points should be considered:
1. $gs_1 \equiv_c gs_1' \Rightarrow \alpha \approx_c \alpha'$
2. $s_1(x) \equiv_c s_1'(x) \Rightarrow mb_{1x} \approx_c mb_{1x}'$
On the basis of the above results (1,2), $gs_2 \equiv_c gs_2'$ can be concluded.

- **Send Statement:** Assume that $s_{1x} = (v_x, mb_{1x}, \langle y.m()\,after\,\gamma|\sigma_x\rangle)$ such that γ_ℓ and γ_r are relative time values and $s_{1y} = (v_y, mb_{1y}, \sigma_y)$. After executing send statement in gs_1, $msg = (x, y, m, \beta)$ will be appended to mb_{1y} such that $\beta_\ell = \alpha_\ell + \gamma_\ell$ and $\beta_r = \alpha_r + \gamma_r$. From $gs_1 \equiv_c gs'_1$, it can be concluded that $s'_{1x} = (v_x, mb'_{1x}, \langle y.m()\,after\,\gamma|\sigma_x\rangle)$ and $s'_{1y} = (v_y, mb'_{1y}, \sigma_y)$ exist in ITTs. After executing send statement in gs'_1, $msg' = (x, y, m, \beta')$ will be appended to mb'_{1y} such that $\beta'_\ell = \alpha'_\ell + \gamma_\ell$ and $\beta'_r = \alpha'_r + \gamma_r$:

$$mb_{2y} = mb_{1y} \oplus msg = mb_{1y} \oplus (x, y, m, \beta)$$
$$mb'_{2y} = mb'_{1y} \oplus msg' = mb'_{1y} \oplus (x, y, m, \beta')$$

Time interval of gs_2 is $\alpha_{r\leftarrow t}$ such that $t = min(\alpha_r, \beta_\ell)$ and Time interval of gs'_2 is $\alpha'_{r\leftarrow t'}$ such that $t' = min(\alpha'_r, \beta'_\ell)$:

$$gs_2 = (s_2, \alpha_{r\leftarrow t}) = (s_1[x \mapsto (v_x, mb_{1x}, \sigma_x)][y \mapsto (v_y, mb_{2y}, \sigma_y)], \alpha_{r\leftarrow t})$$
$$gs'_2 = (s'_2, \alpha'_{r\leftarrow t'}) = (s'_1[x \mapsto (v_x, mb'_{1x}, \sigma_x)][y \mapsto (v_y, mb'_{2y}, \sigma_y)], \alpha'_{r\leftarrow t'})$$

To prove the equivalency of gs_2 and gs'_2, the following points should be considered:
1. $s_2(x) \equiv_c s'_2(x)$
2. $msg \approx_c msg' \wedge mb_{1y} \approx_c mb'_{1y} \Rightarrow mb_{2y} \approx_c mb'_{2y} \Rightarrow s_2(y) \equiv_c s'_2(y)$
3. $\alpha_{r\leftarrow t} \approx_c \alpha'_{r\leftarrow t'}$
On the basis of the above results (1–3), it can be concluded that $gs_2 \equiv_c gs'_2$.

So, in both cases of internal transitions, $gs_2 \equiv_c gs'_2$ as required for proving Eq. 11.

Time Progress Transition: In ITTS, two types of time progress transitions were defined (Eq. 4 and 5). So, it has to be proven that the first condition holds for these two types. In the following, we assume that $gs_1 = (s_1, \alpha)$, $gs'_1 = (s'_1, \alpha')$, $gs_1 \equiv_c gs'_1$.

- **Type 1 Time Progress Transition:** Assume that Type 1 time progress transition is enabled in gs_1. On the basis of (4), $\alpha_\ell < EE_1(gs_1)$ and $gs_1 \overset{TP}{\to} gs_2$. From $gs_1 \equiv_c gs'_1$, it can be concluded that $\alpha'_\ell < EE_1(gs'_1)$. Therefore Type 1 time progress transition is enabled in gs'_1 and $gs'_1 \overset{TP}{\to} gs'_2$. According to Eq. (4), gs_2 and gs'_2 are of the following forms:

$$gs_2 = (s_1, [EE_1(gs_1), EE_2(gs_1)])$$
$$gs'_2 = (s'_1, [EE_1(gs'_1), EE_2(gs'_1)])$$

To prove that $gs_2 \equiv_c gs'_2$, the following points should be considered:
1. $gs_1 \equiv_c gs'_1 \Rightarrow s_1 \equiv_c s'_1$
2. $gs_1 \equiv_c gs'_1 \Rightarrow EE_1(gs'_1) = EE_1(gs_1) + c \wedge EE_2(gs'_1) = EE_2(gs_1) + c \Rightarrow [EE_1(gs_1), EE_2(gs_1)) \approx_c [EE_1(gs'_1), EE_2(gs'_1))$
On the basis of the above results (1,2), $gs_2 \equiv_c gs'_2$ can be concluded.

- **Type 2** Time Progress Transition: Assume that Type 2 time progress transition is enabled in gs_1. On the basis of Eq. (5), $\alpha_\ell = EE_1(gs_1)$ and message (x, y, m, γ) exists in $B(gs_1)$ such that $\gamma_\ell = EE_2(gs_1)$. Therefore $gs_1 \xrightarrow{TP} gs_2$. On the basis of $gs_1 \equiv_c gs_1'$, it can be concluded that $\alpha_\ell' = EE_1(gs_1')$ and message (x, y, m, γ') exists in $B(gs_1')$ such that $\gamma_\ell' = EE_2(gs_1')$. Therefore Type 2 time progress transition is enabled in gs_1' and $gs_1' \xrightarrow{TP} gs_2'$. According to Eq. (5), gs_2 and gs_2' are of the following forms:

$$gs_2 = (s_2, [EE_2(gs_1), EE_3(gs_1)]) = (ds(s_1, EE_2(gs_1)), [EE_2(gs_1), EE_3(gs_1)])$$
$$gs_2' = (s_2', [EE_2(gs_1'), EE_3(gs_1')]) = (ds(s_1', EE_2(gs_1')), [EE_2(gs_1'), EE_3(gs_1')])$$

To prove that $gs_2 \equiv_c gs_2'$, the following points should be considered:
1. $gs_1 \equiv_c gs_1' \Rightarrow ds(s_1, EE_2(gs_1)) \equiv_c ds(s_1', EE_2(gs_1')) \Rightarrow s_2 \equiv_c s_2'$
2. $gs_1 \equiv_c gs_1' \Rightarrow [EE_2(gs_1), EE_3(gs_1)) \approx_c [EE_2(gs_1'), EE_3(gs_1'))$

On the basis of the above results (1,2), it can be concluded that $gs_2 \equiv_c gs_2'$.

Therefore Eq. 11 holds for both types of time progress transitions.

The proof of the second condition of the bisimulation relation (i.e. Eq. 12) is almost the same as the first condition and omitted from this paper because of lack of space.

6   Conclusion

The correctness of behavior in real-time systems, depends on the calculated values and the time of producing theses values [14, 22]. In many real life applications, nondeterministic time behavior is present; In such circumstances, time intervals can be used to define the time behavior of a real-time system.

In this paper, a time interval extension to Timed Rebeca was presented. Timed Rebeca with intervals can be used for modeling real-time systems with nondeterministic time behavior. Using this method, the models of such real-time systems can be described with a high-level language and they can be efficiently analyzed. The semantics of Timed Rebeca with intervals models was defined as Interval Time Transition System (ITTS). In ITTS, every time feature is defined as an interval of non-negative real numbers. A time interval is associated with every system state. The semantics of ITTS was explained and messages, system states, and transitions for different action types were defined. Using the presented semantics, the state space of timed actor systems with time intervals could be generated. In order to determine equivalent system states, shift equivalence relation in ITTS was defined. Using bi-simulation method, it was proved that shift equivalence relation in ITTS could help detect equivalent system states. In many cases with finite time intervals, state space explosion could be prevented using shift equivalence relation in ITTS.

Other equivalence relations can be proposed in the future for detection of equivalent states in cases with infinite time intervals. Another line of research is implementation of the proposed semantics and testing its efficiency on actor models.

References

1. Aceto, L., Cimini, M., Ingólfsdóttir, A., Reynisson, A.H., Sigurdarson, S.H., Sirjani, M.: Modelling and simulation of asynchronous real-time systems using Timed Rebeca. In: Mousavi, M.R., Ravara, A. (eds.) Proceedings 10th International Workshop on the Foundations of Coordination Languages and Software Architectures, FOCLASA 2011, Aachen, Germany, 10th September 2011. EPTCS, vol. 58, pp. 1–19 (2011). https://doi.org/10.4204/EPTCS.58.1
2. Agha, G.A.: ACTORS - a model of concurrent computation in distributed systems. MIT Press Series in Artificial Intelligence. MIT Press, Cambridge (1990)
3. Alur, R., Dill, D.L.: A theory of timed automata. Theoret. Comput. Sci. 126(2), 183–235 (1994)
4. Bengtsson, J., Larsen, K., Larsson, F., Pettersson, P., Yi, W.: UPPAAL—a tool suite for automatic verification of real-time systems. In: Alur, R., Henzinger, T.A., Sontag, E.D. (eds.) HS 1995. LNCS, vol. 1066, pp. 232–243. Springer, Heidelberg (1996). https://doi.org/10.1007/BFb0020949
5. Bjørk, J., Johnsen, E.B., Owe, O., Schlatte, R.: Lightweight time modeling in Timed Creol. In: Ölveczky, P.C. (ed.) Proceedings First International Workshop on Rewriting Techniques for Real-Time Systems, RTRTS 2010, Longyearbyen, Norway, 6–9 April 2010. EPTCS, vol. 36, pp. 67–81 (2010). https://doi.org/10.4204/EPTCS.36.4
6. Clarke, E.M., Emerson, E.A., Sistla, A.P.: Automatic verification of finite-state concurrent systems using temporal logic specifications. ACM Trans. Program. Lang. Syst. 8(2), 244–263 (1986). https://doi.org/10.1145/5397.5399
7. Gomes, L., Fernandes, J.M., Gomes, L., Fernandes, J.M.: Behavioral Modeling for Embedded Systems and Technologies: Applications for Design and Implementation. Information Science Reference (2010)
8. Green, M.: "How long does it take to stop?" Methodological analysis of driver perception-brake times. Transp. Hum. Factors 2(3), 195–216 (2000)
9. Henzinger, T.A., Manna, Z., Pnueli, A.: Timed transition systems. In: de Bakker, J.W., Huizing, C., de Roever, W.P., Rozenberg, G. (eds.) REX 1991. LNCS, vol. 600, pp. 226–251. Springer, Heidelberg (1992). https://doi.org/10.1007/BFb0031995
10. Hewitt, C.: Description and theoretical analysis (using schemata) of PLANNER: a language for proving theorems and manipulating models in a robot. MIT Artificial Intelligence Technical report 258, Department of Computer Science, MIT, April 1972
11. Khamespanah, E., Khosravi, R., Sirjani, M.: An efficient TCTL model checking algorithm and a reduction technique for verification of timed actor models. Sci. Comput. Program. 153, 1–29 (2018). https://doi.org/10.1016/j.scico.2017.11.004
12. Khamespanah, E., Sirjani, M., Sabahi-Kaviani, Z., Khosravi, R., Izadi, M.: Timed Rebeca schedulability and deadlock freedom analysis using bounded floating time transition system. Sci. Comput. Program. 98, 184–204 (2015a). https://doi.org/10.1016/j.scico.2014.07.005
13. Khamespanah, E., Sirjani, M., Viswanathan, M., Khosravi, R.: Floating time transition system: more efficient analysis of timed actors. In: Braga, C., Ölveczky, P.C. (eds.) FACS 2015. LNCS, vol. 9539, pp. 237–255. Springer, Cham (2016). https://doi.org/10.1007/978-3-319-28934-2_13
14. Kopetz, H.: Real-Time Systems - Design Principles for Distributed Embedded Applications. Real-Time Systems Series. Springer, Heidelberg (2011). https://doi.org/10.1007/978-1-4419-8237-7

15. Liu, J.W.: Real-Time Systems. Integre Technical Publishing Co., Inc., Albuquerque (2000)
16. Manolache, S., Eles, P., Peng, Z.: Memory and time-efficient schedulability analysis of task sets with stochastic execution time. In: Proceedings 13th EUROMICRO Conference on Real-Time Systems, pp. 19–26. IEEE (2001)
17. Reynisson, A.H., et al.: Modelling and simulation of asynchronous real-time systems using Timed Rebeca. Sci. Comput. Program. **89**, 41–68 (2014). https://doi.org/10.1016/j.scico.2014.01.008
18. Sirjani, M., de Boer, F.S., Movaghar-Rahimabadi, A.: Modular verification of a component-based actor language. J. UCS **11**(10), 1695–1717 (2005). https://doi.org/10.3217/jucs-011-10-1695
19. Sirjani, M., Jaghoori, M.M.: Ten years of analyzing actors: Rebeca experience. In: Agha, G., Danvy, O., Meseguer, J. (eds.) Formal Modeling: Actors, Open Systems, Biological Systems. LNCS, vol. 7000, pp. 20–56. Springer, Heidelberg (2011). https://doi.org/10.1007/978-3-642-24933-4_3
20. Sirjani, M., Khamespanah, E.: On time actors. In: Ábrahám, E., Bonsangue, M., Johnsen, E.B. (eds.) Theory and Practice of Formal Methods. LNCS, vol. 9660, pp. 373–392. Springer, Cham (2016). https://doi.org/10.1007/978-3-319-30734-3_25
21. Sirjani, M., Movaghar, A., Shali, A., De Boer, F.S.: Modeling and verification of reactive systems using Rebeca. Fundam. Inform. **63**(4), 385–410 (2004)
22. Stankovic, J.A., Spuri, M., Ramamritham, K., Buttazzo, G.C.: Deadline Scheduling for Real-Time Systems: EDF and Related Algorithms. The Springer International Series in Engineering and Computer Science, vol. 460, 1st edn. Springer, New York (2012). https://doi.org/10.1007/978-1-4615-5535-3. https://books.google.com/books?id=1TrSBwAAQBAJ

Path Planning with Objectives Minimum Length and Maximum Clearance

Mansoor Davoodi[ID], Arman Rouhani[✉][ID], and Maryam Sanisales[ID]

Department of Computer Science and Information Technology,
Institute for Advanced Studies in Basic Sciences (IASBS), Zanjan, Iran
{mdmonfared,arman.rouhani,maryamsan}@iasbs.ac.ir

Abstract. In this paper, we study the problem of bi-objective path planning with the objectives minimizing the length and maximizing the *clearance* of the path, that is, maximizing the minimum distance between the path and the obstacles. The goal is to find Pareto optimal paths. We consider the case that the first objective is measured using the Manhattan metric and the second one using the Euclidean metric, and propose an $O(n^3 \log n)$ time algorithm, where n is the total number of vertices of the obstacles. Also, we state that the algorithm results in a $(\sqrt{2}, 1)$-approximation solution when both the objectives are measured using the Euclidean metric.

Keywords: Path planning · Shortest path · Bi-objective optimization · Clearance · Pareto optimality · Approximation algorithm

1 Introduction

Path Planning (PP) is one of the challenging problems in the field of robotics. The goal is to find the optimal path(s) for two given start and destination points among a set of obstacles. However, minimizing the length of the path is usually considered as the optimality criterion. The application of the other objectives of the problem, such as smoothness and clearance has been also considered in the literature [3]. For example, in many applications, the robot needs to move around in order to perform its task properly. The need for moving around the environment, led to the question of what path a robot can take to accomplish its task, in addition to being safe. In this paper, we define the *optimal path* with respect to two objectives minimizing the length of the path and maximizing the clearance (i.e., the minimum distance between the path and the obstacles).

Computing the Visibility Graph (VG) of the obstacles is a classical approach to obtain the path with the minimum length. For a set of polygonal obstacles with n vertices, VG is computed in $O(n^2 \log n)$ time using a ray shooting technique [6]. VG is one of the best-known approaches to obtain the shortest path where the distance between the path and the obstacles is equal to zero – a path with

© IFIP International Federation for Information Processing 2020
Published by Springer Nature Switzerland AG 2020
L. S. Barbosa and M. Ali Abam (Eds.): TTCS 2020, LNCS 12281, pp. 101–115, 2020.
https://doi.org/10.1007/978-3-030-57852-7_8

clearance zero. Hershberger et al. [7] proposed an efficient planar structure for the PP problem in $O(n \log n)$ time.

Clarkson et al. [1] presented two algorithms for finding the rectilinear shortest path amongst a set of non-intersecting simple polygonal obstacles in the plane. One of the algorithms takes $O(n \log^2 n)$ time and the other one requires $O(n \log^{3/2} n)$ time. Likewise, Mitchel et al. [10] used a continuous Dijkstra's algorithm technique which considers the propagation of a "wavefront" for computing the L_1 shortest path among polygonal obstacles in the plane and runs in $O(|E| \log n)$ time and $O(|E|)$ space, where $|E| = O(n \log n)$ is the number of "events". Inkulu et al. [9] presented an $O(n + m(\log n)^{3/2})$ time algorithm to find an $L1$ shortest path, where m is the number of non-intersecting simple polygonal obstacles.

Wein et al. [11] introduced a new type of visibility structure – called *Visibility-Voronoi Diagram* – by using a combination of the Voronoi diagram and VG to find the shortest path for a predefined value λ of clearance. They considered the PP problem in the setting of single objective optimization, that is, minimizing the length subject to minimum clearance λ. Geraerts [5] proposed a new data structure – called *Explicit Road Map* – that creates the shortest possible path with the maximum possible clearance. The introduced structure is useful for computing the path in the corridor spaces. Davoodi [2] studied the problem of bi-objective PP in a grid workspace with objectives minimizing the length of the path and maximizing its clearance. He also studied the problem in the continuous space under the Manhattan metric, and proposed an $O(n^3)$ time algorithm for a set of n vertical segments as the obstacles.

We study the problem of bi-objective PP with the objectives minimizing the length of the path and maximizing its *clearance*, that is, the minimum distance between the path and the obstacles. The goal is computing Pareto optimal solutions, that is, the paths which cannot be shortened if and only if their clearance is minimized. Since this problem is a bi-objective optimization problem in a continuous workspace, there is an infinite number of Pareto optimal solutions. So, it is impossible to provide a polynomial algorithm to compute all the Pareto optimal solutions. To bypass this issue, we focus on different Pareto optimal solutions, the paths with different middle points. We consider a PP workspace with a set of polygonal obstacles with total n vertices; and propose an $O(n^3 \log n)$ time algorithm to compute the Pareto optimal solutions where the length of the paths is measured using the Manhattan and the clearance is measured using the Euclidean metric. Also, we show the algorithm is an efficient approximation approach when both objectives are measured using the Euclidean metric. The problem studied in [2] discusses a special instance of the problem studied here which proposes an algorithm for a set of vertical line segments as the obstacles. Here we study the problem for a set of polygonal obstacles as the obstacles that is a more general case.

This paper is organized in five sections. In Sect. 2, we formally define the problem as bi-objective PP and give some preliminaries. In Sect. 3, an algorithm is proposed for the problem, where the length and clearance objectives are measured using the Manhattan and Euclidean metrics, respectively. The complexity analysis is also in this section. Section 4, extends the results to the case which both objectives are measured using the Euclidean metric and provides an approximation algorithm. Finally in Sect. 5, future work and conclusion are presented.

2   Preliminaries

In this section, we provide some notations to formally define the bi-objective path planning (PP) problem. Let $\mathcal{P} = \{P_1, P_2, ..., P_m\}$ be a set of simple non-intersecting polygonal obstacles, and s and t be the start and destination points in the plane. We denote the set of all vertices of the obstacles by $\mathcal{V}$ and let $n = |\mathcal{V}|$. Also, we denote any collision-free path starting from s and ending at t with the notation $s\text{-}t\text{-}path$. Let $\pi = \, < s = v_0, v_1, ..., v_k, v_{k+1} = t >$ be a rectilinear $s\text{-}t\text{-}$ path containing k $breakpoints$ $v_1, v_2, ..., v_k$, and $\lambda \geqslant 0$ be a favored distance we want to keep away from the obstacles as $clearance$. We define $L(\pi)$ to be the length of π under the Manhattan metric, and $C(\pi) = \lambda$ as the clearance of π under the Euclidean metric, i.e., the minimum distance from the obstacles along π. In the problem of bi-objective PP, we wish to minimize $L(\pi)$ and maximize $C(\pi)$.

Let π_1 and π_2 be two $s\text{-}t\text{-}$paths. We say that π_1 $dominates$ π_2, if π_1 is better than π_2 in both objectives. Precisely, if $L(\pi_1) < L(\pi_2)$ and $C(\pi_1) \geq C(\pi_2)$, or $L(\pi_1) \leq L(\pi_2)$ and $C(\pi_1) > C(\pi_2)$. We also say π_1 and π_2 are $non\text{-}dominated$ paths if none of them dominates the other one. $Pareto$ $optimal$ $paths$ which are denoted by Π^*, are the set of all non-dominated paths in the workspace.

Since Π^* is an infinite set in the bi-objective PP problem, computing and reporting all the Pareto optimal path is impossible. Thus, to have a well-defined problem, we focus on finding the $extreme$ Pareto optimal path defined as follows. Let π_1 be the Pareto optimal path with $C(\pi_1) = \lambda = 0$. By continuously increasing λ, different Pareto optimal paths can be obtained using continuously moving the breakpoints of π_1 in the workspace. This process is possible until λ meets a $critical$ value λ_0. See Fig. 1 to obtain the path π_2 by continuously moving the breakpoints of π_1. In this step, to find a Pareto optimal path π_3 with $C(\pi_3) > \lambda_0$, it is necessary that some breakpoints of π_2 are changed totally (or say jumped), that means, π_3 cannot be obtained using a continuous moving of the breakpoints of π_2 any more. We call π_1 and π_2 as the $extreme$ $Pareto$ $optimal$ $paths$. For example, the subfigures (a)-(d) in Fig. 1 show four extreme Pareto optimal paths π_1-π_4, respectively.

A different point of view for the definition of the extreme Pareto optimal path comes from the $objective$ $space$. Since any $s\text{-}t\text{-}$path π has two objective values $L(\pi)$ and $C(\pi)$, so, it can be mapped to the 2-dimensional objective space $L(\pi)$ and $C(\pi)$. The image of Π^* in the objective space is called $Pareto$ $fronts$. See Fig. 1(e). The extreme Pareto optimal paths are the paths corresponding to the endpoints of Pareto fronts. Indeed, the projection of the Pareto fronts on the axis $C(\pi)$ is a continuous interval $[0, \lambda_{max}]$, where λ_{max} is the maximum possible clearance in the workspace. However, the projection of Pareto fronts on the axis $L(\pi)$ is a set of non-intersected intervals. The lower and upper bound of these intervals correspond with extreme Pareto optimal paths in the workspace.

In the following, we discuss the relationship between the defined bi-objective PP problem and the classic version of the PP problem which is known as shortest path planning problem.

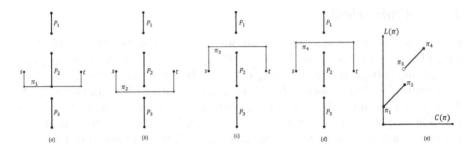

Fig. 1. An example of bi-objective path planning workspace $((a)–(d))$ and its objective space (e). The Pareto optimal paths π_1 and π_2 lie on the same Pareto front, and the Pareto optimal paths π_3 and π_4 lie on another Pareto front. Indeed, the path π_2 can be obtained by continuously moving the breakpoints of π_1 by increasing λ, however, the path π_3 cannot be obtained.

In the classic single-objective version of the problem, the goal is to minimize $L(\pi)$. Therefore, only one path is required as the output. As mentioned earlier, there are several studies that solve the problem of finding the rectilinear shortest path among polygonal obstacles. Although this version is single-objective, it is the same as our problem when $C(\pi) = \lambda = 0$. Precisely, one can solve the bi-objective path planning problem by considering it as many shortest path problems at each desired clearance λ. Instead of keeping a distance λ from obstacles, we can grow (or say fatten) the obstacles with size λ. If we are given a polygon P, the Minkowski sum of P and a disk $\mathcal{D}(\lambda)$ of radius λ is the set of all points whose distance from P is less than λ.

Let $\mathcal{P}(\lambda) = \{P_1(\lambda), P_2(\lambda), ..., P_m(\lambda)\}$ be the set of fattened polygons with $\mathcal{D}(\lambda)$. It is obvious that the breakpoints of the shortest s-t-path for a given clearance λ belong to $\mathcal{P}(\lambda)$ [2]. Therefore, we can increase λ slightly from 0 to $+\infty$, and perform a shortest path algorithm in each level to obtain all extreme Pareto optimal s-t-paths. However, since the problem is considered in continuous space, it is not clear how much we should increase λ to find a new extreme Pareto optimal path. The idea is to determine *events* which make changes in the shortest path during the continuously increasing λ process. Assume we perform a shortest path algorithm when $\lambda = 0$, and let π_1 be the resulted path with some breakpoints. Now we continuously increase λ until at least one of the breakpoints of π_1 changes in λ_c and a new path π_2 is generated. Regarding the definition, π_2 is an extreme Pareto optimal solution. Precisely, when $\lambda = \lambda_c$, the Pareto front changes and a jump to another Pareto front occurs. That is, to obtain a Pareto optimal solution with clearance more than λ_c, it is needed that some breakpoints of π_2 are changed. Therefore, the number of such extreme Pareto optimal paths is equal to the number of these jumps.

3   Bi-objective PP Problem Among Polygonal Obstacles

In this section, we propose an algorithm to find the Pareto optimal solutions and analyze the complexity of this method. In the first subsection we describe the algorithm which is based on all the critical events that may change the shortest s-t-path. We discuss constructing data structures to handle these events. Finally, at the end of this section, we analyze the complexity of the algorithm.

3.1   The Algorithm

Given $\mathcal{P} = \{P_1, P_2, ..., P_m\}$ – m polygonal obstacles with total n vertices – as the input, we are supposed to find all extreme Pareto optimal paths. In order to find the shortest path for a specific clearance λ, a modified version of the Dijkstra's algorithm is used – called SP(λ,$\mathcal{V}$,$\mathcal{P}(\lambda)$). The inputs of this algorithm are the clearance value λ, the set of obstacles' vertices $\mathcal{V}$, and the set of fattened obstacles $\mathcal{P}(\lambda)$. Let T be the constructed tree by this algorithm. Whenever the SP algorithm is performed, T is initialized with the root s and the closest vertices to it as its children. Every node v in T is assigned a weight $w(v) > 0$ that shows the shortest distance between s and v, and has pointers to its parent and children. Obviously, the breakpoints of an s-t-path belong to the set of vertices $\mathcal{V}$ if $\lambda = 0$. It is clear when λ increases, the shortest s-t-path gets tangent to the fattened obstacles. For a polygonal obstacle P including a vertex v_i, let $P(\lambda)$ be the fattened obstacle and $TP_{v_i}(\lambda) = \{v_i^r(\lambda), v_i^l(\lambda), v_i^u(\lambda), v_i^d(\lambda)\}$ be the set of most four main directions, or say *tangent points*, of vertex v_i moving as a function based on λ (see Fig. 2 for an illustration). Since the shortest s-t-path is rectilinear, it can only be tangent to the $P_i(\lambda)$ at the tangent points of the vertices of it. For example, in Fig. 2, the path can be tangent to v_i at $v_i^l(\lambda_\epsilon)$, $v_i^u(\lambda_\epsilon)$, or both of them. An s-t-path touches at most three *tangent points* of a vertex (see Fig. 3).

Observation 1. *The tangent points of every vertex v_i can be computed as a linear function based on λ.*

Proof. Let $P(\lambda)$ be the fattened obstacle containing v_i as its vertex. Since this fattened obstacle is obtained by performing a Minkowski sum with the disc $\mathcal{D}(\lambda)$, the *tangent points* of v_i lie on the boundary of $\mathcal{D}(\lambda)$ in each of the four directions with the distance λ from v_i. Thus, the coordinates of the tangent points can be easily computed as follows:

$$\begin{cases} y_{v_i^r}(\lambda) = y_{v_i} \\ x_{v_i^r}(\lambda) = x_{v_i} + \lambda \end{cases} \qquad \begin{cases} y_{v_i^l}(\lambda) = y_{v_i} \\ x_{v_i^l}(\lambda) = x_{v_i} - \lambda \end{cases} , \qquad (1)$$

$$\begin{cases} y_{v_i^u}(\lambda) = y_{v_i} + \lambda \\ x_{v_i^u}(\lambda) = x_{v_i} \end{cases} \qquad \begin{cases} y_{v_i^d}(\lambda) = y_{v_i} - \lambda \\ x_{v_i^d}(\lambda) = x_{v_i} \end{cases} . \qquad (2)$$

$\square$

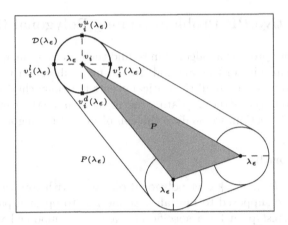

Fig. 2. Fattened obstacle $P(\lambda_\epsilon)$ and the *tangent points* of vertex v_i ($TP_{v_i}(\lambda_\epsilon)$).

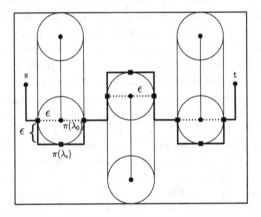

Fig. 3. The shortest s-t-path $\pi(\lambda_0 = 0)$ is shown in dotted chain and the shortest s-t-path $\pi(\lambda_\epsilon)$ is shown in solid bold chain.

The SP algorithm first creates $TP_{v_i}(\lambda)$ (for $0 \leqslant i \leqslant n$) according to Eqs. (1) and (2). Next, it performs the Dijkstra's algorithm on the resulted tangent points to construct T. Note that the four mentioned tangent points of every vertex become equal to their origin when $\lambda = 0$. After performing the SP algorithm, all the tangent points are assigned a positive weight. Mitchell [10] proposed an appropriate planar subdivision method based on segment dragging techniques that can be used to find the adjacency relations between the nodes. This part is adopted from [10]; however, to make the paper self-contained, we provide some basic notations here.

Let $S = (\theta, \Phi_1, \Phi_2)$ be a subdivision where a sweeping line is at inclination θ and it is being dragged parallel to itself whose endpoints lie on the rays l_1 and l_2 which have inclinations Φ_1 and Φ_2 respectively. By building the subdivisions $S = (3\pi/4, 0, \pi/2)$, $S = (5\pi/4, \pi/2, \pi)$, $S = (7\pi/4, \pi, 3\pi/2)$, and $S = (\pi/4, 3\pi/2, 2\pi)$, the closest points from any node under the L_1 metric are obtained. Therefore, for $O(n)$ vertices, the adjacent vertices can be found in $O(n \log n)$ time and $O(n)$ space.

Let π be the shortest path reported by the SP algorithm when $\lambda = 0$. As mentioned earlier, this path remains optimal until at least one of its breakpoints changes while λ increases. According to the definition, the new path π' is an extreme Pareto solution and it is accepted as a new optimal path. The following lemma shows that the change in the parent node of one of the breakpoint nodes can result in a new path.

Lemma 1. *During increasing λ, the optimal path does not change unless a change in the parents of some nodes in T occurs.*

Proof. Based on the *principle of optimality*, any sub-path of a shortest s-t-path is also a shortest path. So, let π be the current shortest path and π' be the new shortest path which is generated by the SP algorithm at some critical clearance λ_c. For the sake of contradiction, suppose that π' is obtained without any changes in T. Thus, if we backtrack π', the parents of the breakpoints remain unchanged. Apparently by moving backward from t to s in π, the parent node of each node is the closest node (based on the weights in the Dijkstra's algorithm) to it, and when no parent changes, this means the shortest path remains the same and this contradicts π' to be the new shortest s-t-path.   □

According to this lemma, we have to keep track of changes in T (because it stores the parent and child relations and when a change in these relations occurs, T changes accordingly). Let $T(\lambda)$ be the tree at clearance λ. As mentioned previously, these changes are called *events*. We propose an algorithm to check all such events in $T(\lambda)$. We also construct data structures to handle them. The same as what is done in [2] and [11] every changes in T should be checked to find possible new paths. And it is obvious that not more than three types of events (where T changes) may occur while λ increases:

- *Type1:* A tangent point reaches a Manhattan weighted Voronoi diagram edge.
- *Type2:* Two tangent points reach the same x or y coordinate.
- *Type3:* Two obstacles intersect at some clearance λ_I.

In the following, we discuss these events. Let $MWVD(\mathcal{V})$ be the Manhattan weighted Voronoi diagram of V. This diagram partitions the plane into the regions each corresponding to a point p. Each point q that lies inside the region of p, is closer to point p compared to the other points. For the region corresponding to point p that contains all the points q. Thus, we have:

$$\forall p' \in V, p' \neq p \Rightarrow d(p,q) + w(p) \leq d(p',q) + w(p'), \tag{3}$$

where $d(.,.)$ is the Manhattan distance between two points and $w(p)$ is the weight of point p. if p be the parent of p', we have:

$$w(p') = d(p,p') + w(p). \tag{4}$$

After performing the SP algorithm, a positive weight $w(v)$ is assigned to each node which represents the length of the shortest rectilinear path from s to v. According to the Eq. (4), the weight function of each node is a linear function that can be computed based on λ and the weight of its parent as an inline function. Note that tangent points move based on λ. The $MWVD(\mathcal{V})$ is constructed in $O(n \log n)$ time [8]. This diagram has some properties compared to the Euclidean Voronoi diagram. All the edges in the $MWVD(\mathcal{V})$ are horizontal, vertical or with a slope of $\pm\pi/4$. As λ increases, these edges move as a function of λ. For simplicity, edges between two points p and q and their equations as a function of λ are presented as follows.

Let $Rect(p_\lambda, q_\lambda)$ be the rectangle cornered at $p_\lambda(x_p - \lambda, y_p - \lambda)$ and $q_\lambda(x_q + \lambda, y_q + \lambda)$ (see Fig. 4(a)) or $p_\lambda(x_p + \lambda, y_p - \lambda)$ and $q_\lambda(x_q - \lambda, y_q + \lambda)$ (see Fig. 4(b)). Figure 4 shows two different positions of the two points p and q. For reasons of symmetry, we consider only one side and the lines on the other side can be computed easily with the same method. The equations are as follows:

$$\begin{cases} Y_{L_1} = -1/2x_q + 1/2((y_p + y_q) + x_p - (w(p) - w(q))), \\ X_{L_1} = x_q + \lambda, \end{cases} \tag{5}$$

$$Y_{L_2} = -X_{L_2} + 1/2(2\lambda + (x_p + x_q) + (y_p + y_q) - (w(p) - w(q))), \tag{6}$$

$$\begin{cases} Y_{L_3} = 1/2x_p + 1/2((y_p + y_q) - x_q - (w(p) - w(q))), \\ X_{L_3} = x_p + \lambda, \end{cases} \tag{7}$$

$$Y_{L_4} = X_{L_4} + 1/2(-2\lambda - (x_p + x_q) + (y_p + y_q) - (w(p) - w(q))). \tag{8}$$

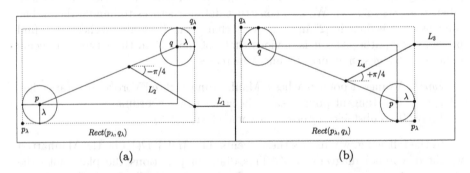

(a)   (b)

Fig. 4. Two different positions of the two points p and q and their $MWVD(\mathcal{V})$ edges.

Corollary 1. *The $MWVD(\mathcal{V})$ edges L_1 and L_2 (or L_3 and L_4) of two vertices $p(x_p, y_p)$ and $q(x_q, y_q)$ (see Fig. 4) intersect the $Rect(p_\lambda, q_\lambda)$ at the same point.*

Whenever a Type1 event occurs, an $MWVD(\mathcal{V})$ edge is passed. Of course, it is possible that a Type1 event occurs and no parent changes. After constructing $MWVD(\mathcal{V})$, each vertex lies inside a cell of Voronoi diagram and the parent of each vertex is known. Critical λs are the intersection points of a vertex v with the edges of $MWVD(\mathcal{V})$ between the parent of v and its adjacent vertices. A Type1 event is shown in Fig. 5. The parent of the vertex p is pp before the event. When the intersection takes place, it means that the parent vertex should be changed to the new vertex qq in which p is going to enter its Voronoi region.

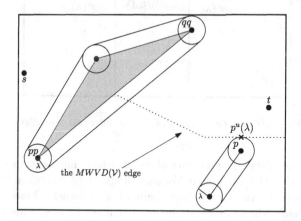

Fig. 5. A *Type*1 *event* between the fattened obstacles.

When two tangent points reach the same x or y coordinate, a Type2 event occurs. For simplicity, we introduce some definitions in order to find such critical λs. Let $H(v_i)$ and $V(v_i)$ be the horizontal and vertical lines passing through vertex v_i (for $0 \leqslant i \leqslant n$), respectively. Each $H(v_i)$ and $V(v_i)$ ends either at the first obstacle edge to which they are incident or at the vertex v_i (if it is not tangent to the obstacle that contains v_i). The lines with the latter characteristic, do not move when λ increases. Let $H_{V_i}(\lambda)$ and $V_{V_i}(\lambda)$ be the moving horizontal and vertical lines based on λ:

$$H_{V_i}(\lambda) = C_y + \lambda, \tag{9}$$

$$V_{V_i}(\lambda) = C_x + \lambda, \tag{10}$$

where C_y and C_x are the x and y coordinates of the points they pass through for $\lambda = 0$. Since this event occurs when two vertical or horizontal lines intersect, we can compute the critical λ at which they reach the same coordinate at a constant time. A Type2 event is shown in Fig. 6. The parent node of $p^u(\lambda)$ is $pp^u(\lambda)$ before two moving lines $H_q(\lambda)$ and $H_{pp}(\lambda)/H_p(\lambda)$ reach the same y coordinate. When the this event occurs, $q^b(\lambda)$ becomes the parent of $p^u(\lambda)$.

The Type3 event occurs when two polygons intersect – called *intersection events*. For simplicity, assume that polygons do not intersect simultaneously and just two polygons intersect at each intersection event. Precisely, during increasing λ, when the polygons get fattened, they may intersect at some critical λs. When intersection events take place, the parents of some nodes can be changed (see

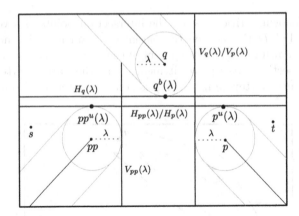

Fig. 6. A *Type2 event* between the fattened obstacles.

Fig. 7 for an illustration). The critical λs at which intersection events occur, can be found with the Euclidean Voronoi diagram of $\mathcal{P}$. When two obstacles intersect, we unite them and consider them as one obstacle. Thus, each time an intersection occurs, the number of obstacles reduces by one. This type of event may change the parent of some nodes. A Type3 event is shown in Fig. 7. When $\lambda = 0$, the parent of v_4 is v_3 and the parent of v_5 is v_4. When the obstacles intersect, since the previous path is closed, the tangent point $v_5^b(\lambda)$ becomes the parent of $v_4^b(\lambda)$. The total complexity of the Euclidean Voronoi diagram of polygons, is $O(n)$ and it can be constructed in $O(n \log n)$ time [4,11]. Since the Voronoi diagram of the polygons contains $O(n)$ edges [4], there are $O(n)$ critical λs at which polygons may intersect. Let $\mathcal{I} = \{I_{\lambda_1}, I_{\lambda_2}, ..., I_{\lambda_{O(n)}}\}$ be the set of the critical λs related to the intersection Type3 events.

Observation 2. *Since all the relations in T might be changed in the worst case when a Type3 event occurs, the problem of bi-objective path planning can be decomposed into $O(n)$ different clearance intervals. In each interval, the tree structure T is reconstructed, the weight function of each tangent point is computed, and the Type 1 and Type2 events can be handled.*

Any two consecutive members of the ordered set $\mathcal{I}$ construct an interval. More precisely, $[I_{\lambda_1}, I_{\lambda_2}), [I_{\lambda_2}, I_{\lambda_3}), ..., [I_{\lambda_{O(n)-1}}, I_{\lambda_{O(n)}}), [I_{\lambda_{O(n)}}, +\infty)$ are the set of such intervals. According to the observation 2 we consider each interval separately, and perform the following steps at the beginning of each interval until s and t are no more in the same connected component.

First, we perform the SP algorithm on the tangent points in order to construct T and obtain the weight of each node. As aforementioned, the weight of each node is a linear function that is obtained based on λ and the weight of its parent. Note that once each node gets weighted, the $MWVD(\mathcal{V})$ is obtained based on Eqs. (6)–(9) accordingly. It is obvious that the weight of the origin vertices can be obtained from the weight of tangent points according to Eqs. (2) and (3) in order to compute the $MWVD(\mathcal{V})$.

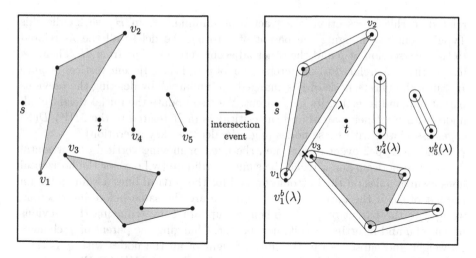

Fig. 7. The intersection of two obstacles that makes changes in T.

Second, all the critical λs that cause a change in T related to Type1 and Type2 events are computed.

Lemma 2. *The intersection of a tangent point and an edge of $MWVD(\mathcal{V})$ (Type1 event) is efficiently computable.*

Proof. A Type1 event occurs when a tangent point intersects an edge of $MWVD(\mathcal{V})$. For simplicity, we consider horizontal and vertical lines tangent to the obstacles at tangent points. When an obstacle intersects an edge of $MWVD(\mathcal{V})$, these lines and the edges intersect. Since all the moving lines – fixed lines and the edges of $MWVD(\mathcal{V})$ – are based on linear functions, the intersection points and such critical λ can be found in constant time for each event.   □

Lemma 3. *The intersection of two horizontal or vertical lines associated to each vertex (Type2 event) is efficiently computable.*

Proof. As mentioned earlier, each vertex has two vertical and horizontal tangent lines. Depending on the obstacles and the location of the vertices, these lines can be moving or fixed. In both cases, the related equations are linear. Therefore, the critical λ, at which these lines reach the same x or y coordinate, can be found in constant time for each event.   □

Third, we manage the data structures used to handle the events that make changes in T. Let H be the heap structure by which we store critical λs related to Type1 events and Λ be the sorted list of critical λs associated with Type2 events. In each interval, the next critical λ, which should be handled, is the minimum λ that is extracted from H and Λ.

Assume p is a tangent point that intersects an $MWVD(\mathcal{V})$ edge between adjacent nodes pp and qq, and let pp be the parent of p. When the critical λ

related to this intersection is extracted as minimum from H, we handle this Type1 event by changing the parent of p to qq. We delete all the λs related to the intersection of p and the other adjacent vertices of pp from H. Then, we insert critical λs related to the intersection of p and the adjacent vertices of qq to H. Since the parent node of p is changed, we update T by deleting the previous parent of p and inserting its new parent. We also update the initial weight of all nodes with p as their parent and the critical λs in H related to the $MWVD(\mathcal{V})$ edges associated with these nodes (e.g., the decrease key operation).

When a Type2 event occurs, i.e., the fixed or moving vertical or horizontal lines of two nodes p and q reach the same coordinate (w.l.o.g. for the horizontal lines assume p lies on the bottom of q and for the vertical lines assume p lies on the left of q). If the parent nodes of p and q are the same before intersection, we update the parent of q to p. Then, we update T by removing the previous parent of q and inserting p as its new parent. Changing the parent of q, changes its weight and subsequently the initial weight of all the nodes with q as their parent. Thus, the critical λs in H related to the edges of $MWVD(\mathcal{V})$ associated with these nodes should be updated sequentially.

3.2   Complexity Analysis

In this subsection we discuss the complexity of the proposed algorithm. In order to find the intersection events, we construct the Voronoi diagram of polygons under the Euclidean distance in $O(n \log n)$ time. Since $|\mathcal{I}| = O(n)$, the complexity of each interval is multiplied by $O(n)$.

Remark 1. The Dijkstra's algorithm runs in $O(n \log n + m)$ using a Fibonacci heap, where m is the number of edges of the graph. Since the graph is planar in our problem, m equals to $O(n)$ and the Dijkstra's algorithm runs in $O(n \log n)$ time.

Lemma 4. *The Type2 events can be handled in the worst case $O(n^3 \log n)$ time.*

Proof. Every vertex has a horizontal and a vertical tangent line. Thus, in the worst case, $O(n^2)$ events of this type can occur over all the intervals. When an event of such type occurs, it takes $O(n)$ time to update T and the weight of its children. We also update the edges of $MWVD(\mathcal{V})$ related to the nodes whose weights have been updated which cause updates in H. Let N_p be the number of neighbors of node p. The process of updating each edge of $MWVD(\mathcal{V})$ between the two nodes, takes a constant time. Updating the heap structure for each vertex p also takes $O(N_p \log n)$. So,

$$A = \sum_{p \in \mathcal{V}} (N_p \log n) = \log n \sum_{p \in \mathcal{V}} N_p = O(n \log n).$$

Therefore, the total complexity time is $O(n^2 * n \log n)$, that is $O(n^3 \log n)$.   □

Lemma 5. *The Type1 events can be handled in the worst case $O(n^3 \log n)$ time.*

Proof. According to the proposed algorithm, we insert the intersection point of a node to all of the $MWVD(\mathcal{V})$ edges corresponding to its parent adjacent nodes to H as critical λs. As the number of edges is $O(n)$, the total number of adjacent nodes is $O(n)$ as well. Thus, similar to lemma 4, it can be proved that the total complexity time for updating H is $O(n \log n)$ in each interval which is $O(n^2 \log n)$ in total. The same as lemma 4, we update T and the edges of $MWVD(\mathcal{V})$ that are between the nodes whose weights have been updated. By updating the edges of $MWVD(\mathcal{V})$, the heap structure H should be updated which can entirely be done in the worst case $O(n^2 \log n)$ for all events in each interval. Thus, the total complexity is $O(n^3 \log n)$. $\qquad\square$

Therefore, we can conclude this section with the final result as follows.

Theorem 1. *The problem of bi-objective path planning among a set of polygonal obstacles with the objectives minimizing the length and maximizing the clearance, where the length and the clearance are measured using the Manhattan and the Euclidean metrics, respectively, can be solved in $O(n^3 \log n)$, where n is the total number of obstacles' vertices.*

4   An Approximation Algorithm for Euclidean Bi-objective Path Planning

Consider the bi-objective PP problem with the objectives minimizing the length and maximizing the clearance under the Euclidean metric. We denote this problem with EbPP. Before going to a formal definition for an approximation for a bi-objective optimization problem, we denote that EbPP is a challenging open problem which follows some questions as:

1. How many extreme Pareto optimal paths are there in the worst case?
2. Does the problem of EbPP belong to the class of *Polynomial* problems?
3. What is the lower bound on the EbPP problem?

Based on our studies on this problem, when the length of the paths is measured using the Euclidean metric, more complicated algebraic weighted functions are replaced with the simple weighted functions presented in the previous section for the Manhattan case. So, we believe that the hardness of the problem of EbPP is beyond the class of the polynomial problems. However, we guess the number of the extreme Pareto optimal solutions is $O(n^3)$ as well as the conjecture $\Omega(n^2)$ as the lower bound of the problem, where n is the number of total vertices of the polygonal obstacles. Therefore, as an initial approach, we present an approximation solution for the problem of EbPP.

The following definition for an (α, β)-*approximation* Pareto optimal solution is adopted from [2].

Definition 1. *Let Π be a bi-objective minimization problem with the objectives f_1 and f_2. A solution X is an (α, β)-approximation Pareto optimal solution for Π, if there is no solution Y such that $f_1(X) \geq \alpha f_1(Y)$ and $f_2(X) > \beta f_2(Y)$, or $f_1(X) > \alpha f_1(Y)$ and $f_2(X) \geq \beta f_2(Y)$.*

Remark 2. The proposed algorithm in the previous section which is optimal for the bi-objective path planning problem when the length of the path is measured using the Manhattan metric, provides $(\sqrt{2}, 1)$-approximation extreme Pareto optimal path for EbPP as well.

5   Conclusion

In this paper, we considered the problem of bi-objective path planning with the objectives minimizing the length and maximizing the clearance. We assumed a general case of the problem in the plane where the obstacles are a set of simple polygons with n vertices in total. We proposed an $O(n^3 \log n)$ time algorithm for finding all extreme Pareto optimal solutions of the problem where the length and clearance of the paths are measured using the Manhattan and Euclidean metrics, respectively. We also showed that such paths are good approximation solutions when both objectives are measured using the Euclidean metric. However, proposing an efficient algorithm to find the Pareto optimal paths for the latter problem remains a hard open problem as mentioned in Section four.

References

1. Clarkson, K., Kapoor, S., Vaidya, P.: Rectilinear shortest paths through polygonal obstacles in $O(n(\log n)^2)$ time. In: Proceedings of the Third Annual Symposium on Computational Geometry, pp. 251–257 (1987)
2. Davoodi, M.: Bi-objective path planning using deterministic algorithms. Robot. Auton. Syst. **93**, 105–115 (2017)
3. Davoodi, M., Panahi, F., Mohades, A., Hashemi, S.N.: Clear and smooth path planning. Appl. Soft Comput. **32**, 568–579 (2015)
4. de Berg, M., van Kreveld, M., Overmars, M., Schwarzkopf, O.: Computational geometry. In: Computational geometry, pp. 1–17. Springer, Heidelberg (1997). https://doi.org/10.1007/978-3-662-03427-9_1
5. Geraerts, R.: Planning short paths with clearance using explicit corridors. In: 2010 IEEE International Conference on Robotics and Automation, pp. 1997–2004. IEEE (2010)
6. Ghosh, S.K., Mount, D.M.: An output-sensitive algorithm for computing visibility graphs. SIAM J. Comput. **20**(5), 888–910 (1991)
7. Hershberger, J., Suri, S.: An optimal algorithm for Euclidean shortest paths in the plane. SIAM J. Comput. **28**(6), 2215–2256 (1999)
8. Hwang, F.K.: An O(n log n) algorithm for rectilinear minimal spanning trees. J. ACM (JACM) **26**(2), 177–182 (1979)
9. Inkulu, R., Kapoor, S.: Planar rectilinear shortest path computation using corridors. Comput. Geom. **42**(9), 873–884 (2009)

10. Mitchell, J.S.: L_1 shortest paths among polygonal obstacles in the plane. Algorithmica **8**(1–6), 55–88 (1992). https://doi.org/10.1007/BF01758836
11. Wein, R., Van den Berg, J.P., Halperin, D.: The visibility-Voronoi complex and its applications. Comput. Geom. **36**(1), 66–87 (2007)

Author Index

Printed in the United States
by Baker & Taylor Publisher Services

Printed in the United States
by Baker & Taylor Publisher Services